Dispatches from the People's War in Nepal

'This unique, intimate look into the People's War in Nepal provides invaluable background to the world's most vigorous Maoist movement, and insight into the theory and practice underlying contemporary Maoism elsewhere in South Asia and globally. Based on the author's reportage and interviews in guerrilla-controlled areas in 1999, *Dispatches from the People's War in Nepal* helps to explain why, five years later, the insurgency has acquired control over most of the Nepali countryside.'

Gary Leupp, Professor of History at Tufts University and Coordinator of the Asian Studies Program

'In her dispatches from the ongoing revolutionary war in Nepal, where she was the first, and longest-staying, foreign journalist to report from the Maoist-held areas, Li Onesto keeps up the committed, conscientious revolutionary journalism of John Reed, George Orwell, and Agnes Smedley. Building around the narratives of guerrilla soldiers and their families, of group leaders, farmers, local officials, teachers, and artists, she provides an intimate and sympathetic view of the early stages of the People's War while giving a sense of the arduous nature of fighting a war in the Himalayas. Hers is probably the best, if not only, account of how the Maoists built their organization and movement, and of how they operate and govern.'

Stephen Mikesell, author of *Class, State and Struggle in Nepal: Writings 1989–1995*

'This lively, exciting and enlightening presentation of the true portrait of the Maoist insurgency in Nepal will help people to understand the real state of affairs behind the "People's War" waged by the Communist Party of Nepal (Maoist) to liberate the Nepalese people from all kinds of exploitation and repression. The most important value of this book lies in its serious analysis of several human features of the Maoist Revolution with on-the-spot descriptive facility.'

Padma Ratna Tuladhar, independent left leader, senior human rights leader and one of the facilitators in the peace talks between His Majesty's Government of Nepal and the Communist Party of Nepal (Maoist). Served as the Minister of Health and Labor in the Nepalese government, member of the Nepalese Parliament, chairman of the Forum for Protection of Human Rights, and founder of the 'Save Democracy Movement'.

Dispatches from the People's War in Nepal

Li Onesto

Pluto Press
LONDON • ANN ARBOR, MI

and

insight
Insight Press, Inc.
CHICAGO, IL

First published 2005 by Pluto Press
345 Archway Road, London N6 5AA
and 839 Greene Street, Ann Arbor, MI48106
www.plutobooks.com
and
Insight Press, Inc.
4064 N. Lincoln Ave., #264, Chicago, IL 60618

British Library Cataloguing in Publication Data
A catalogue record for this book is available from the British Library

ISBN 0 7453 2341 3 hardback (Pluto)
ISBN 0 7453 2340 5 paperback (Pluto)
ISBN 0 9760236 0 1 paperback (Insight)

Library of Congress Cataloging in Publication Data applied for

10 9 8 7 6 5 4 3 2 1

Designed and produced for Pluto Press by
Chase Publishing Services, Fortescue, Sidmouth, EX10 9QG, England
Typeset from disk by Stanford DTP Services, Northampton, England
Printed and bound in the European Union by
Antony Rowe Ltd, Chippenham and Eastbourne, England

Crisis & Opportunity

This is the Chinese phrase,

"Opportunity in the Midst of Crisis."

The Chinese word for "crisis" has two parts:

"danger" 危 and "opportunity" 机 ,

and the character 中 means "center."

Contents

List of Photographs

Acknowledgements

Many people helped make this book possible.

A special thanks goes to my editor Sylvia Alexis for her dedication to this project from the very beginning. Before my trip, she helped me develop an approach to this story. When I returned she read hundreds of pages of my notes and critiqued my photos with her usual artistic rigor. Most importantly, she worked with me to construct the narrative and analysis of *Dispatches from the People's War in Nepal*. She went through many drafts, and was always there with fresh ideas, encouragement, and friendship. We have been through a real journey together to produce a book that can bring this little known, but highly important struggle to the attention of people around the world.

I have benefited enormously from the work of Bob Avakian. In preparing to witness a Maoist people's war, I studied his book *Mao Tsetung's Immortal Contributions*. And in thinking through and trying to understand the strategic problems confronting the revolution in Nepal, I have found his writings on Maoism and the world historic challenges of the communist revolution invaluable. His insights have provided a sweeping historical framework, not only for this project but also for my continuing study and analysis of developments in Nepal as well as other parts of the world. In one of the most remote corners of the earth, I was excited and heartened to see Maoist guerrillas reading the works of Bob Avakian.

In the course of conceptualizing, researching, and writing this book, I have gained greatly from the suggestions and insights of fellow journalists, colleagues, and friends. Particular thanks go to my fellow writers at the *Revolutionary Worker* newspaper who have always been there with critical thinking, original ideas, and enthusiasm, and to Lisa Rivers for her meticulous and informed proofing. Thanks also to Raymond Lotta for his comments and input which helped hone the story and analysis. I am highly appreciative of the support and enthusiasm given by Roger van Zwanenberg at Pluto.

Several people have provided technical and artistic assistance with my photos. These photos have been an important part of this story and have enabled many people to see the human face of this revolution. I am indebted to JT and ES for their hard work on my photography exhibits and on the design of the photo section of this book.

Finally, I am greatly indebted to all those in Nepal who helped make my trip possible. The Communist Party of Nepal (Maoist) gave me the unique opportunity to travel as a photo-journalist into the guerrilla zones and allowed me access to political and military leaders, as well as to students, women, and peasants involved in mass revolutionary organizations. I thank the family members of people killed in the war, who came to tell me their painful stories in the hopes that, as a result, the world will know about their struggle. I am grateful to the many peasants who opened their doors to us on our travels, sometimes in the wee hours of the morning, offering us food, shelter, and conversation. And a special thanks goes to all those who translated for me while I was in Nepal, especially 'Pravat,' who not only provided a 24-hour essential link to all the people around me in Rolpa and Rukum but also gave his friendship – and along with others, kept me from falling off steep cliffs as we trekked in the dark of night.

The revolution in Nepal is generating important political and theoretical analysis, and in working on this book I have drawn from writings by Prachanda, the leader of the CPN (Maoist), as well as work by Baburam Bhattarai and Parvati.

As the reader will soon discover, I have been deeply inspired by the energy, consciousness, and dedication of the masses of people in Nepal fighting to bring a new world into being. Ultimately, this book owes its existence to them.

1996-2004: Eight Years of People's War in Nepal

February 13, 1996	The Communist Party of Nepal (Maoist) initiates armed struggle against the Nepalese government with simultaneous attacks in different areas of the country. Approximately 5,500 large- and small-scale actions are carried out in the following weeks.
May 1998	The government launches Kilo Sera 2, a major counterinsurgency operation, transferring tens of thousands of specially trained armed police forces to 20 of the country's 75 districts. In the first two months, at least 200 people are killed.
May 1999	K.P. Bhattarai is elected prime minister in nationwide elections boycotted by the Maoists.
March 2000	Prime Minister Bhattarai is ousted by Girija Prasad Koirala, who accuses Bhattarai of failing to maintain law and order. Koirala promises to use all possible means, including deployment of the Royal Nepal Army (RNA), to combat the Maoists.
September 25, 2000	About 1,000 guerrillas raid the headquarters of the Dolpo District in Dunai. Fourteen policemen are killed. Prime Minister Koirala wants to send in the RNA in pursuit of the Maoists but the army brass refuses. Home Minister Govinda Raj Joshi resigns in protest.
February 2001	Parliamentary opposition parties demand Prime Minister Koirala's resignation, citing his implication in a bribery scandal with Lauda Air, and his inability to maintain law and order.

April 2001	Koirala recommends sending the RNA to fight the Maoists in some areas. With the king's consent, Koirala deploys the RNA to protect development projects in seven rural districts as part of a government-run Integrated Security and Development Package.
May 2001	Maoists hold mass rallies to announce the formation of local revolutionary governments in their regional strongholds in the west.
June 1, 2001	King Birendra and eight other members of the royal family are murdered in the Royal Palace. The king's brother, Gyanendra, takes the throne under a cloud of suspicion. Different political forces demonstrate in response to the killings. A curfew is declared and hundreds are arrested.
June 2001	The CPN (Maoist) helps form CCOMPOSA (Coordination Committee of Maoist Parties and Organizations of South Asia), made up of ten parties and organizations, with the stated purpose to 'unify and coordinate activities of Maoist parties and organizations in South Asia to spread protracted people's war in the region.'
July 6–13, 2001	Maoist guerrillas attack police posts in Lamjung, Nuwakot, Ramechhap, Gulmi, Dailekh, and Holeri. RNA military forces stationed nearby do not go in to rescue 69 captured policemen.
July 19, 2001	Prime Minister Koirala announces his resignation. The new prime minister, Sher Bahadur Deuba, calls for a unilateral ceasefire and negotiations with the Maoists.
July–September 2001	The Maoists agree to a ceasefire and talks. Three meetings are held between the CPN (Maoist) and the Nepalese government. The Maoists put forward their demands for an interim government, a constituent assembly, a new constitution, and a republic. They also

demand the release of their people in prison. A number of Maoists, including some top leaders, are freed.

September 2001 After the September 11 attacks in the US, the Nepalese government expresses support for the US 'war against terrorism.' Government officials in India call the Maoists in Nepal 'terrorists' and pledge support for the government of Nepal.

The CPN (Maoist) formally announces the formation of the People's Liberation Army.

November 21, 2001 Prachanda, chairman of the CPN (Maoist), says there is no reason for further talks with the government because the CPN's main demands have been refused.

November 23, 2001 Maoist guerrillas launch a new offensive, carrying out actions in more than 20 of the country's 75 district headquarters.

The CPN (Maoist) announces a new 37-member 'United People's Revolutionary Council' to govern the areas where the Maoists now have control.

The People's Liberation Army (PLA) now has several permanent companies, and in some cases is fighting in units of brigade strength – of several hundred soldiers.

King Gyanendra, for the first time, unleashes the RNA to fight the Maoists. A State of Emergency ushers in months of arrests, censorship, and suspension of constitutional rights. India offers more weapons, and the US, China, Russia, and Japan issue statements supporting the king in his efforts to defeat the Maoists.

January 17, 2002 For the first time in 30 years, a high US official visits Nepal. Secretary of State Colin Powell meets with the king, prime minister, and top military officials.

May 2002	Prime Minister Deuba meets with US President George W. Bush in Washington, DC and requests political, financial, and military help to defeat the Maoists. Deuba also goes to Europe to ask for help. Britain provides $40 million and the US comes up with $22 million for the Nepalese regime. The US sends a dozen military experts to survey Nepal and map out operational plans for the RNA.
June 2002	The British government hosts an international meeting to discuss how different countries can help the Nepalese government defeat the Maoist insurgency.
October 4, 2002	King Gyanendra, in essentially a palace coup, removes Prime Minister Sher Bahadur Deuba, assumes executive power, and dissolves the entire Council of Ministers. Gyanendra indefinitely postpones elections scheduled for November and appoints a new prime minister, Lokendra Bahadur Chand, of the Rastriya Prajatantra Party (RPP), a pro-monarchist party.
November–December 2002	Amnesty International reports that nearly half the people killed in the People's War have been civilians targeted for their real or perceived support of the Maoists. The death toll since the start of the war stands at 7,000.
January 29, 2003	A new ceasefire and round of negotiations begins. The government agrees to withdraw the 'terrorist' label it had formally placed on the Maoists and cancels the international arrest warrants issued through Interpol and the bounties placed on Maoist leaders.
May 2003	The US State Department puts the CPN (Maoist) on one of its official lists of 'terrorist' organizations.

	During the negotiations, the US provides military aid and training to the RNA and delivers 5,000 M-16 rifles with the promise of 8,000 more.
August 17, 2003	As a third meeting of talks is taking place, RNA soldiers murder 19 Maoists in a village in the eastern district of Ramechhap. According to Amnesty International, security forces opened fire on a house, one Maoist was killed and the other 18 were taken away and executed one by one.
August 27, 2003	Prachanda issues a statement exposing the 'cold-blooded killings' by the RNA and the government's refusal to seriously discuss the Maoists' main demands. Fighting resumes.
October 2003	The US government declares the CPN (Maoist) a threat to US national security and freezes the group's assets as part of a package of sanctions. CPN (Maoist) announces they have control of 80 percent of the rural areas in Nepal.
February 13, 2004	The CPN (Maoist) celebrates the eighth anniversary of the People's War.

Nepal

Credit: *A World to Win Magazine*

Introduction

June 1, 2001, King Birendra and Queen Aishworya are with their family at the Narayanhiti Royal Palace in Kathmandu, Nepal.

Suddenly, around 10:40 pm, the regular Friday night dinner turns into bloody carnage.

Crown Prince Dipendra had left dinner early, but now returns, drunk and dressed in military fatigues. He sprays the room with a semi-automatic rifle, then shoots himself in the head with a pistol.

Most are dead on arrival at the hospital – the king and queen, two of their children, and six other members of the royal family have been killed.

According to the constitution, the murderer, Prince Dipendra, is the new king and although brain dead is kept alive in the hospital.

Birendra's brother, Prince Gyanendra, is conveniently out of town on this fateful night. He's next in line to the throne and is declared king after Dipendra is taken off life support.

No guards or aides were in the room to witness the killings and few believe the official story that 29-year-old Dipendra, educated at Britain's Eton College, suddenly went berserk because his mother disapproved of the woman he wanted to marry.

Thousands gather in the streets of Kathmandu, many suspicious that someone in the government is behind the massacre. Police use batons to disperse a crowd of 1,000 people who throw stones at the police and shout slogans against the prime minister.

President Bush and Pope John Paul II send condolences to Gyanendra. India declares three days of state mourning. Queen Elizabeth and the Prince of Wales are said to be 'deeply shocked and saddened,' and royal palaces, residences, and government buildings in Britain are ordered to fly flags at half-mast.

* * *

This bizarre tale of regicide – like something out of *King Lear*, with a Columbine, automatic weapons twist – made international headlines. All of a sudden, newspaper articles were running stories about the social, political, and economic situation in Nepal. And in a strange way, this was how many people around the world learned, for the first time, about the Maoist revolution that has been going on in Nepal since 1996.

People who had simply thought of Nepal as 'home to Mount Everest' and a great place to trek and see beautiful scenery were now reading about a guerrilla insurgency that was growing, gaining popular support, and establishing control over much of the countryside.

Most news accounts dutifully reported the official story of the palace massacre – ruling out any political motivation behind the incident. One US diplomat called it the result of an 'incredible quarrel in the family that went incredibly bad.'[1] But more astute observers knew that this palace bloodbath took place in the context of, and was very much linked to, intense political disputes within Nepal's ruling class.

In June 2001, the defining political question causing huge debate and crisis among Nepal's rulers was the same as it is today – how to deal with the Maoist insurgency. Birendra had been criticized by many for not moblilizing the Royal Nepal Army against the Maoists. But King Gyanendra quickly proved less reluctant to do this. Before the end of that year, the RNA had been fully unleashed against the guerrillas.[2]

* * *

In the spring of 1999, I had the unique opportunity to travel into the guerrilla zones of Nepal. I stayed in villages where poor farmers provided food and shelter. I traveled and lived with members of the People's Army and interviewed political and military leaders, guerrilla fighters, relatives of those killed in the war, and villagers in areas under Maoist control. I embarked on this three-month journey with the aim of capturing the passion, voices, and faces of the peasants who are waging what they call a 'People's War.' I was the first foreign journalist to be given such access by the Communist Party of Nepal (Maoist) – the party leading this revolution. And I came back with notebooks full of interviews, hundreds of photos, and tapes of music performed by guerrilla 'cultural squads.' *Dispatches from the People's War in Nepal* chronicles my journey and is the first book that provides this kind of up close, inside human story of the revolution in Nepal.

Nepal is one of the poorest and most underdeveloped countries in the world. Living conditions are extremely primitive, even by Third World standards. Per capita income is less than $200 a year and some 70 percent of the population live below the poverty line. There is extreme class polarization and social inequality. Ten percent of the population earn 46.5 percent of the national income and own 65 percent of the cultivable land; 85 percent of the population live in the rural areas, most without electricity, running water, and basic sanitation. There are hardly any doctors in the countryside and malnutrition is widespread. Life expectancy is only 55 years. The infant mortality rate is more than 75 per 1,000, about ten times the rate of Japan and Sweden. According to government data, the literacy rate is less than 50 percent. Most areas are so rugged and isolated that it takes days of trekking across steep terrain to get to where people live.[3]

The history and conditions leading up to the Maoist insurgency in Nepal are unique and fascinating. Nepal was never formally colonized but, since the early 1800s, has been subject to foreign domination, especially by Britain and India. Today, with a history of extreme dependence on India, Nepal has almost no industry. Like many poor African countries, Nepal remains under the dominance of the world market but has not been the object of the large-scale sweatshop-style investment that has marked 'globalization' in many other Third World countries.

Nepal is sometimes thought of as a peaceful Shangri-La. But in fact, from the time it was unified in the late 1700s, there has been a history of armed struggle against the government and foreign domination.

In 1815, the Nepalese people waged guerrilla warfare against Britain and in many places defeated the British army. But the Nepalese monarchy surrendered to England and Nepal was forced to give up a third of its territory.

Throughout the 1800s, peasant rebellions frequently broke out against dictatorial regimes. And after the Second World War, discontent against the government intensified as many people saw the monarchy and the ruling elite as agents of Indian domination. In the midst of growing social conflict, a Nepalese communist party was formed in 1949.

In the early 1950s thousands of Indian troops were brought in to put down a major rebellion in western Nepal. But peasants, in some places led by communists, continued to defy the government, and the next several decades were marked by armed peasant rebellions.

From 1951 to 1990, Nepal was ruled by the Shah monarchy. All political parties were outlawed under the feudal panchayat system in which the country was run by councils – panchayats – and the king was the ultimate authority. Then, after widespread protest in 1990, known in Nepal as the 'Janodalon Uprising,' the king was forced to institute a constitutional monarchy with an elected parliament. General elections were held for the first time in 1991. But hopes that the new parliamentary system would bring progressive change were soon dashed. An unstable government – nine prime ministers in the first ten years – has been unable to solve the country's deep social and economic problems and is widely seen as beholden to India and thoroughly corrupt.

Indian domination has been a major and longstanding factor in Nepal's political history. The monarchy, as well as the Nepali Congress (the main parliamentary party today), have had longstanding ties with and backing from the Indian power structure. In 1950–51, India directly intervened to put King Tribuvan on the throne. Throughout the 1960s, India provided support for the Nepali Congress forces waging armed struggle against the King Mahendra government. And during the 1990 Janodalon uprising, India worked behind the scenes to push its own agenda and exert political pressure on the various political forces that were in rebellion against the absolute monarchy.[4]

India has long considered Nepal strategically important in its often hostile relationship with China. Conflict between India and China intensified after the victory of the Chinese Revolution in 1949 and this impacted on India's demands on Nepal. In 1965, aiming to prevent friendly relations being established between Nepal and the socialist government in China, India secured an arms treaty which stipulates that Nepal must purchase arms only from India, Britain, and the US, or other countries recommended by India. Today, this treaty remains in force.[5]

The Maoists in Nepal have denounced the current Chinese government as 'revisionist' – socialist in name, but capitalist in fact. And the post-Mao regime in China has 'disowned' the guerrillas in Nepal and shares India's hostility to this Maoist revolution. But India remains concerned that China could take advantage of the instability created by the insurgency to expand its power and influence in the whole region.

It is largely, though not exclusively, through India that Nepal is linked to the world capitalist system. India dominates the economic life of the country – plundering Nepal's natural resources, enforcing unequal trade agreements, and exploiting the Nepalese peasants who cross the border looking for work.

India obtains raw materials such as timber from Nepal, along with massive amounts of cheap hydroelectric power. In 1996, the Mahakali Treaty basically established India's right to steal Nepal's water. While Nepal is one of the poorest countries in the world, its mountains and rivers make it one of the richest in water resources – third in the world after Brazil and China. Nepal has as much capacity to generate hydroelectricity as the US, Mexico, and Canada combined. But unequal treaties force Nepal to sell much of its water to India at give-away prices. Meanwhile, 40 percent of the rural population in Nepal lacks regular supplies of potable water and only about 10 percent of the country have access to electric power.[6]

Nepal also provides India with a close market for goods. A 1950 'peace and friendship treaty' between the two countries has prevented Nepal from establishing and developing a national industry. According to this treaty, all industrial production needed by the Nepalese people is to be supplied from India.[7]

Every year, huge waves of poor farmers migrate to India in search of work. In this way, millions of Nepalese peasants become part of the working class. Some stay for years; others work for several months and then return home to farm. Ironically, this has created a favorable factor for the revolution in Nepal. There is a strong movement of Nepalese Maoists in India who support the People's War in Nepal. And there are friendly ties between the Maoists in Nepal and Indian Maoists waging armed struggle against the Indian government. I met people in Nepal who were first exposed to Maoism while working in India – and then returned home to join the guerrillas.

This particular history in Nepal – combined with the influence of international events and political trends – has created conditions for a strong Maoist movement.

Nepal is not the only country where Marxist parties have long been a significant part of the political scene – including within the government. But what is interesting here is that a distinctly *Maoist* movement has gained widespread influence, especially among the poor peasantry, and that after initiating armed struggle against the government in 1996, its strength has continued to grow. The Maoists in Nepal are going against all the official verdicts that have declared socialist- and communist-led movements no longer relevant or viable in today's world. But within five to six years of starting armed struggle, the Communist Party of Nepal (Maoist) had won extensive popular support and gained control of most of the countryside.

The generation of activists in Nepal who fought against the monarchy in the 1960s and early 1970s were greatly influenced by the revolution and the establishment of socialism in China, Maoist movements in India, and anti-colonial struggles erupting around the world. Prachanda, the chairman of the CPN (Maoist), is part of this generation. In an extensive interview I did with him on my trip, he told me, 'When the Great Proletarian Cultural Revolution was initiated in China under the leadership of great comrade Mao, it directly impacted on the revolution in Nepal. There were so many materials from the Chinese Cultural Revolution that came to Nepal. This Cultural Revolution inspired mainly the younger generation of communists and the masses.'[8]

Even after Mao's death in 1976 and the reversal of his politics in China, there continued to be a large movement in Nepal inspired by his ideology and military doctrine. Today Prachanda still upholds the relevance of Mao's vision that what is needed in a country like Nepal is a 'New Democratic Revolution' aimed at overthrowing the current regime and going on to establish a new socialist society.

Mao's theory holds that revolution in the oppressed countries passes through two stages. The first stage is the new-democratic revolution. This is not a bourgeois-democratic revolution that leads to the establishment of a capitalist system, but a revolution led by the proletariat aimed at decisively breaking the grip of imperialism on the country and deeply transforming the social system and eliminating the pre-capitalist economic and social relations – especially the survivals of feudal or semi-feudal relations. In this stage, it is necessary and possible to build a broad united front of all classes and strata that can be united to overthrow imperialism, feudalism and bureaucrat-capitalism, under the leadership of the proletariat and its party. In Maoist theory, this revolution clears the way for the second stage, the establishment of socialism.

The CPN (Maoist) makes a point of emphasizing that their revolution is part of the worldwide struggle for communism and they are a participant in the Revolutionary Internationalist Movement – which is made up of parties and organizations around the world that uphold 'Marxism-Leninism-Maoism.'[9] According to Maoists the new-democratic revolution, from the beginning, must be carried out with a clear strategic perspective of socialism and communism. So while there are stages in this process, it is conceived of as a unified process – guided throughout by the outlook, ideology and politics of the proletariat and its goal of a communist world. There is a red thread running from protracted people's war as the road for carrying out the new-democratic revolution all the way through to the establishment of a socialist society and the continuation of the revolution under the dictatorship of the proletariat, all as part of the advance of the world revolution.

Among the many communist parties in Nepal, there has always been sharp debate over whether to work within the multiparty system or to take up arms to overthrow it. So it was a defining moment when the CPN (Maoist) made their big move in 1996. In an international political climate where guerrilla movements in South Africa and El Salvador had already dropped their guns to seek elected positions and reforms, the Maoists in Nepal denounced the 'parliamentary road' and initiated armed struggle aimed at overthrowing the government.

In the mid-1990s, Prachanda and other leading members of the CPN (Maoist) analyzed that the conditions in Nepal were ripe for launching, building, and sustaining an armed struggle and that such a struggle could unite and mobilize Nepal's peasantry. The specific characteristics of Nepal which they saw as a basis for launching and winning a 'people's war' included: a corrupt, semi-feudal, centralized state system with little reach outside the main cities; geographic and demographic factors favorable to revolutionary mobilization including the fact that large parts of the population live in remote areas where the government presence is weak; deep poverty; the widespread influence of communism among the people; and large numbers of Nepalese working in India, who, exposed to revolutionary ideas, could act as a kind of rear support area.[10] The CPN (Maoist) knew that the looming presence of India would pose a real threat to a revolution in Nepal, but also recognized that there is great revolutionary potential throughout the whole area, with Nepal a potential example for revolution throughout the whole subcontinent.

In early 1995, the Maoists began a year-long campaign to build support among the peasants for initiating war. Centered in the western districts of Rolpa, Rukum, and Jajarkot, the Maoists sent political-cultural teams into the villages, organized the peasants to challenge local authorities and mobilized villagers to build roads, bridges, and latrines.

From the beginning, the CPN (Maoist) conceived of their revolution as a 'protracted war' – the type propounded by Mao Tsetung as applicable in semi-feudal and semi-colonial countries. Mao's basic theory of 'protracted people's war' recognized that in semi-colonial, semi-feudal countries like Nepal, the revolutionary forces start out weak and small compared to the government forces, and that to engage in all-out military battles would only lead to getting crushed. But by avoiding decisive tests of strength and waging guerrilla warfare, the revolutionary forces can defeat and weaken the government forces in smaller battles and through a protracted process gain popular support, increase in strength and numbers, and extend their control. Building rural base areas and establishing military control and political authority in ever larger parts of the countryside allows the revolutionaries to surround the cities from the countryside, and eventually seize country-wide political power.[11]

The decision to embark on such a path in Nepal was very controversial in the revolutionary movement – as well as within the Party itself. Many significant leaders of the communist movement in Nepal had argued that there could be no successful revolutionary war in Nepal – that the country was encircled, that a revolution could only be crushed, and that it would be essentially foolhardy to start such a war. Prachanda told me, 'In making the plan for initiation there was great debate over how to go to the armed struggle because many people were influenced by "peaceful" struggle, work in the parliament, rightist and petty bourgeois feelings, and a long tradition of the reformist movement. Then we said that the only process must be a big push, big leap. Not gradual change.'[12]

The Maoists recognized that starting a war with the government would be difficult – fraught with uncertainty and risk. Before 1996, many of the leadership and cadre in the CPN (Maoist) were working underground. But initiating armed struggle and developing a people's army required a whole new level of commitment. At the time armed struggle began, what the Party leadership calls the 'Initiation,' the Party was mainly made up of educated intellectuals. Some CPN (Maoist) leaders recounted to me how some of these cadres did not agree with the decision to launch armed struggle and others were reluctant to leave their jobs, go underground, and become 'full-timers.' This led some Party members, even some who had been in leading positions, to leave the Party. Others quit because they could not withstand the new levels of government repression.

In the first few years of fighting, as the government launched a series of counterinsurgency campaigns, many veteran Party members were killed, and new, younger leaders had to step into their shoes. Throughout this process, the character of the Party changed as more peasants were recruited. Today, the Party and the People's Liberation Army are overwhelmingly made up of peasants.

I spent a month traveling through the Rolpa and Rukum districts in the Western Region – which have been and remain the key areas of strength for the Maoists. These areas are extremely remote and far from the seat of power in Kathmandu – and rapid and large mobilization of government forces to these areas is difficult. Even before 1996, Maoist forces had a lot of influence in these poor districts where the Magar people – one of the 25 or so oppressed nationalities in Nepal – make up much of the population. Most of the guerrillas I met in these areas were Magars, attracted to the Maoists' promise to end discrimination based on caste and nationality and uphold the right of self-determination for oppressed ethnic groups.

Official statistics portray Nepal as a 'Hindu country.' But in Rolpa and Rukum many of the people are minority nationalities that do not practice the Hindu religion and I was struck by the absence of Hindu temples. One guerrilla told me that the lack of strong religious influences in this part of the country had made it easier for revolutionary ideas to take hold.

Many different sections of Nepalese society have been drawn to support the Maoists' fight against the government: the rural population wants land and development, women want equality and an end to oppressive feudal and patriarchal traditions, the minority nationalities want an end to discrimination and the cruel caste system, and millions throughout Nepal want democratic rights and national independence. Broad sections of the people have come to support this Maoist-led revolution as providing the way to achieve these things as part of the overall revolutionary transformation of society.

There are also many middle-class forces that are sympathetic to the demands of the Maoist revolution. In Kathmandu, I met intellectuals, artists, and even high-level government workers who supported the Maoists' program of ending semi-feudal despotism and foreign domination. But the heart of this revolution is in the countryside and it is here that the guerrillas have built their base areas of power.

Within the first few years of the revolution, government officials, landlords, and police had been driven out of many villages in Rolpa and Rukum. This has created a power vacuum, which has allowed the guerrillas to establish military, political, and economic control. The development of such 'base areas' is a strategic part of the Maoists' plan to eventually 'surround the cities and seize power.' Today, millions of people live in such areas where the guerrillas are constructing the foundations of the new society they hope to build if they succeed in overthrowing the government.

* * *

The following story of my trip to Nepal in 1999 provides an in-depth picture of the roots and beginnings of the People's War in Nepal.

What are the conditions that have provided the basis and fertile ground for the rapid growth of this revolution?

How did this Maoist-led war get underway and begin to take hold among larger sections of the population?

The crucial question of land comes through in conversations with poor farmers who describe their struggle to eke out a living on small plots of land. Stories of endemic corruption and cruelty by local officials, moneylenders, and landlords illustrate why millions in the countryside have lost faith in the current regime. Young women describe what it is like to face feudal traditions that suffocate and deny them equality from the day they are born – and why they became guerrilla fighters. Peasants from lower castes and oppressed minority nationalities describe why they have placed their hope for equality with the Maoist revolution.

Military commanders and high-level Party leaders – including top leaders in the Party's Western Regional Bureau, members of the Central Committee's Politbureau, and the leadership of mass organizations – discuss the first few years of the insurgency: the days and months after the Initiation of armed struggle – how the people's army and militias evolved from primitive fighting groups into trained squads and platoons; the growing involvement of women; and the first government campaigns aimed at crushing the small but growing insurgency. They also recount the year before the war was launched – how the Party carried out various campaigns to build support for going over to armed struggle and to prepare the Party politically, militarily, and organizationally to take this dramatic step.

While many people around the world have been unaware of the developing conflict in Nepal, many more eyes are now focused on this situation, especially as the United States, Britain, and other major world powers have moved to directly intervene in the situation by giving military, financial and political support to Nepal's counterinsurgency efforts.

The US has allocated millions of dollars in aid to Nepal, supplied thousands of machine guns and other weaponry, and provided military advisers and training for the RNA. In 2002, Britain hosted an international meeting to discuss how different countries could help the Nepalese regime defeat the Maoists. At the meeting, the UK's

Foreign Office minister, Mike O'Brien, stated that Nepal's struggle against the insurgents should be seen as part of the wider 'war against terrorism.' Later, during a visit to Kathmandu, O'Brien said, 'The clear message of my visit is this: that the Maoists will not be allowed to win here in Nepal; they cannot be allowed to win.'

After September 11, 2001, the Nepalese government, along with India, labeled the Maoists in Nepal 'terrorists.' The US increasingly began to talk about the conflict in Nepal in the context of its 'war against terrorism.' A proposal initiated by President Bush for $20 million in economic and military aid to Nepal stated:

> We currently do not have direct evidence of an al-Qaeda presence in Nepal, but weak governance has already proved inviting to terrorists, criminals and intelligence services from surrounding countries ... continued instability in Nepal could create the conditions in which terrorists easily could establish operations, especially in remote areas in the far west of the country ...[13]

Behind such tortured efforts to compare the Maoists in Nepal to groups like al-Qaeda are real concerns by the US over the geostrategic implications of a Maoist victory in a region of the world that is already tremendously volatile and of considerable geopolitical significance.

Today, the Nepalese regime, Western government officials and authorities, and intelligence analysts all acknowledge that in a relatively short period and up against major counterinsurgency campaigns by the government, the Maoists have gained the support of millions of peasants, are waging successful guerrilla warfare against government forces, and have established political control in most of the countryside.[14] All this makes the story of the origins and beginnings of this People's War all the more fascinating. One should remember that at the time the CPN (Maoist) launched their revolution it was not at all determined that they would get as far as they have. In fact, it was with a good amount of confidence – as well as brutal determination – that the Nepalese government responded with their campaigns aimed at quickly eradicating the guerrillas. This book's journey back to the beginning years of the People's War in Nepal sheds light on why the Maoists in Nepal have been able to go from a relatively small political party to a party leading an army and political apparatus that is effectively exercising power in most of Nepal's countryside and is now in a position to make a serious attempt to seize power in Kathmandu.

When people ask me why they should be interested in the People's War in Nepal, I point to the state of the planet.

We live in a world where the richest countries have 20 percent of the world's people but 85 percent of its income. We live in a world where 840 million people are chronically malnourished and more than ten million children die each year as a result of preventable diseases.[15] This is a world in which women's very humanity is taken away by veils, chadors, and burkhas; where women are exploited and brutalized by sweatshops, a booming sex trade, and domestic violence. This is a world in which capitalist globalization, heralded as the wave of the future and the roadmap to prosperity, is in fact intensifying inequality and suffering. Today, peasant farmers in the Third World, who still make up half of humankind, are being ruined, dispossessed, and uprooted from the land on a scale never before witnessed in history.

These are immense problems. And clearly, big solutions are called for.

But is it possible to achieve truly egalitarian and emancipating change? Or does the choice really come down to *jihad* or McWorld – fundamentalism or consumerism? The people fighting in the battlefields and liberated areas of Nepal believe a different future is possible.

Today, the conventional wisdom of mainstream discourse is that 'communism is dead,' and the People's War in Nepal is often portrayed as an anomaly. Marxist-inspired revolution has been declared a 'failed project' by bourgeois analysts. And even among many radical thinkers, the idea of a 'popular liberation struggle' aimed at seizing power through armed struggle is frequently seen as a relic of the 1960s that has proved unrealistic and misconceived. Yet here in Nepal a popular revolution has emerged that is applying Maoism – in its military tactics, strategies, political vision, ideological orientation, and stated aims and goals. Far from being the 'last gasp' of a dying ideology, the Maoist revolution in Nepal is causing surprise with its success and raising once again the question of the relevance of communist revolution to today's turbulent world.

Li Onesto
June 2004

I
Meeting the People's Army

February 1999: For an hour and 45 minutes I stare out the window of the plane, completely transfixed. Up above the clouds, level with our flight path, lies the immense and amazing Himalayan Range. The white peaks, rising up beyond scattered clouds, look unreal and I feel as if I've been transported to another planet. The pushed-up earth soars skyward and onward with unbroken continuity and wholeness, while each individual mountain has a unique shape and character – some jutting up alone, others clumped together, some white-capped with snow, others wind-swept steel-gray. They stretch far into the blurry distance.

This is my entry to the 'roof of the world': Nepal – famous for Mount Everest, Gurkha soldiers, and Sherpa mountain people – a country revered and raved about by tourists as a great place to hike, take in beautiful scenery, and relax. As for myself, I am planning on hiking and seeing a lot of the countryside. But I'm not planning on doing any relaxing. I am on my way to observe, first hand, a *revolution* going on in the 'foothills' of the Himalayas.

Back in the United States, I had heard about Maoist guerrillas who had been waging armed struggle against the government since 1996. On February 13 of that year, they had reportedly carried out hundreds of actions around the country and proclaimed this as the start of a 'People's War.' Modeled after the strategy developed by Mao Tsetung in China, their plan was to concentrate guerrilla warfare and build base areas in the vast countryside, surround the cities and eventually seize nationwide power.

In response, the Nepalese government launched major campaigns to try and stop the insurgency. According to a 1998 US State Department report, 'Police reaction to the People's War insurgency led to incidents of unwarranted force against prisoners and noncombatants,' and 'police committed numerous human rights abuses.' I had read that more than 600 guerrillas and villagers had been killed by the government in the first three years of fighting. And various human rights organizations had spoken out against the Nepalese police carrying out summary executions, rape, and torture. Yet most people around the world have heard little, if anything, about this conflict.

And what has been in the news, by and large, has been one-sidedly sympathetic to the viewpoint of the Nepalese government, with little information about the goals and character of the insurgents.

The purpose of my trip is to get an up-close look at the human faces and minds of the people fighting this People's War. I want to travel into the very heart of this conflict. I want to interview the leaders and military commanders of the Communist Party of Nepal (Maoist), the party leading this revolution – to find out who they are and why they have resorted to a violent revolution. I want to talk and live with villagers in the war zones – to learn about their conditions of life and investigate the Maoists' claim that they have widespread support among the peasants. And I want to learn, first hand, about the human rights violations being carried out by the government.

Most of all, I hope to get a living, breathing, portrait of this revolution – people describing in their own words the dangerous thinking and passion behind their decision to pick up arms against the government. I want to look into the different elements of Nepalese society that seem to be fueling this revolution – the poverty of peasant farmers, the feudal treatment of women, the oppression of ethnic minority groups, the hopelessness of youth.

There is plenty of news and commentary about the 'Maoist problem' in the mainstream media in Kathmandu. But my mission is to help the world to hear the story of the side in this war that doesn't have easy access to TV, radio, and international newswires. My notebooks and rolls of film are waiting to capture this story.

* * *

When I arrive in Nepal's capital city of Kathmandu, things are very intense. The parliamentary elections, scheduled for May 1999, are shaping up to be a focal point of struggle between the Nepalese government and the People's War. From reading the daily editorials in the *Kathmandu Post*, it seems that the government is hoping to use the elections to project an image of strength, stability, and democracy. And they need to consolidate a parliament that will be firmly behind its efforts to crush the Maoists. But the CPN (Maoist) has called for a countrywide boycott of the elections. In the countryside, guerrilla attacks on police and politicians are on the rise and there have been actions by the revolutionaries in the capital city as well.

In Kathmandu, I get in contact with some members of the CPN (Maoist) to talk about traveling to the guerrilla zones and interviewing people. They are eager to help; they want the chance to have their side of the story told. They tell me that no other foreign journalist has been given extensive access to the guerrilla zones – so I will be the first!

One morning I finally get the word that arrangements have been made for my first trip to the countryside. We're going to the Eastern Region, to an area where the People's War is strong. I get only a couple of hours' notice and am told to pack a small bag since we'll be leaving the city on motorcycle.

In the afternoon I meet up with Shiva, who will be my guide and translator, and we hook up with two men who will take us east. Riding by motorcycle out of Kathmandu is only the first of many death-defying experiences I will have in Nepal. As a pedestrian I am already quite familiar with the hectic and dangerous city traffic. But now, after many days of dodging taxis and motorcycles on crowded streets, I am on the other side of the collision equation, zig-zagging in and out between people, brushing up against other zooming vehicles, and dodging lumbering cows and skittish goats.

As we leave Kathmandu I start to see, for the first time, what the eastern countryside looks like. Soon three- and four-story buildings disappear and give way to smaller brick structures along an increasingly winding and ascending road. In the dusk I see the outlines of irregular green steps wrapped around steepness – amazing terracing done by farmers up and down the mountainside.

The temperature is very pleasant and even with the wind whipping around us on the motorcycle I only need a light windbreaker. It's quickly getting dark but there are still people working in the fields and many people walking along the road. We pass several groups of people celebrating, and the colorful clothing of special occasions whizzes through my peripheral vision and my ears catch snatches of lively music. The driver of the motorcycle tells me there are lots of wedding celebrations going on because this is the traditional month in Nepal for people to get married.

We stop along the road, and even though the sun has completely surrendered behind the mountains, some kids are climbing and playing on the steep paths leading up the hillside. We wait for a while, anxiously. But for some reason we've missed the people we're supposed to hook up with. It's dark and we'll have to wait until morning to try and reconnect. So we decide to go to a small nearby inn for the night. After a standard Nepali meal of *dal bhat* (lentils and rice) we retire to cubbyhole rooms. I'm pretty tired. But I'm also keyed up from the anticipation and excitement of being on my way to my first meeting with guerrillas from the People's Army and I can't fall asleep.

It's 9:30 pm, but outside, across the street, a small workshop is still open and a man is busily beating metal into water jugs. The steady rhythm of steel against steel goes on late into the night, mingling with sounds of people talking and walking on the street. We are near a major road with constant traffic, and throughout the night the noisy honking of buses and trucks intrudes on my restless slumber.

Early in the morning I wake up to the lilt of a man with a beautiful voice singing as he prepares food in the restaurant downstairs. Soon after this, the metal worker across the street starts up again with his rhythmic banging: it's barely after 6:00 am. When I walk out to the back balcony of the inn, I see that we are next to a big river. All around me are immense mountains stretching into the distance.

While I savor the scenery and sip some traditional Nepali milk-tea, I think about the adventure – and danger – ahead of me. People who have arranged my trip to the east have warned me that because of the coming elections this is an especially dangerous time to travel in areas where the People's War is going on. The Royal Nepal Army has not yet been mobilized against the guerrillas, but the police have already carried out many brutal counterinsurgency campaigns. We will travel carefully, but there are no guarantees we won't encounter the police. And they tell me that if the police find out that Maoists are in the villages we are visiting, they might encircle the area and launch an attack.

We have to enter the guerrilla zone after dark, so we arrange to rendezvous with our contacts in the early evening. We are met several miles up river from the inn, just as the sun is setting. We set out on foot for a couple of hours of uphill climbing. Some daylight is still hanging around when we stop to rest along the way and some village kids quickly surround us, their faces full of curiosity.

We arrive at a village in an area the Maoists have designated to be developed towards becoming what they call a 'base area' – where the government is no longer in control and a new revolutionary authority can be established.

People here say support for the People's War is so strong in this area that the police are afraid to enter for fear they will be killed. Guerrilla sentries protect the surrounding territory and if police come anywhere near, people wanted by the police and people working underground are notified and leave. The Maoist leaders in this area know we are coming and make sure things are secure for our arrival.

It's very dark by the time we get to the house where we'll be staying and meeting with people. We immediately go inside and are greeted by two of the local leaders of the CPN (Maoist). One of them, Rohit,* is a teacher and we will be staying at his house. The other is a young man, Surya, who looks to be in his early twenties. We sit down and over milk-tea they tell me something about themselves and the revolutionary work in this village. Someone brings in a kerosene lamp so I can write in my notebook. I notice there is wiring on the walls for electricity and even sockets for light bulbs. Apparently the house was built in the hope that someday power lines would reach this area. But this village, like 90 percent of Nepal, still has no electricity.

Surya tells me that people in this village are dependent on farming and grow mainly corn and millet. Most of the peasants have very small plots of land which yield small harvests, so they are only able to grow enough food to feed their families for three or four months. The rest of the year they have to find some other means of income to survive. Some have small gardens where they grow crops such as tomatoes, which can be sold in the cities. Many of the men are forced to leave their families for months at a time to find work elsewhere.

Surya is the son of a carpenter. Two of his brothers have gone to live and work in Kathmandu, and two of his sisters have joined the revolution. After finishing high school he went to Kathmandu to study law, got involved in the revolutionary student movement, and joined the Party in 1994. He tells me, 'Then the Party wanted me to do work in the countryside and, at the same time, government repression forced me to leave the city and go underground.'

Rohit has been a teacher for 20 years. He used to teach in the city, but more than ten years ago he returned to the village. He tells me his wife is also involved in the revolution. He has six brothers – one is in India working as an oil worker, another drives a truck in the city. His mother and father live in this house with his wife and children.

Surya and Rohit are leaders in an area of about 2,000 people. The Maoists have been organizing here for ten years, and before the start of the People's War in 1996 – what the Party calls 'the Initiation' – there were already established party cells (units of party members)

* Most names of people I interviewed have been changed to protect their identity, except the families of those who have been killed.

and revolutionary mass organizations in this area. In Nepal there are a number of other 'communist parties.' But the CPN (Maoist) considers them 'revisionist' – Marxist only in name, while carrying out reformist, or in some cases reactionary, pro-government politics. One such party, the Communist Party of Nepal (United Marxist Leninist), commonly referred to as the UML, actually headed the Nepalese government for nine months in 1994–95 and used to do a lot of political organizing in this area. But I'm told that many people here who have been UML members and supporters in the past now support the People's War, are neutral, or have moved away. Rohit fills in this picture with some of his own experience with the UML. He says:

> 'I joined UML when I was a student in 1980 and was involved in the student movement, following the politics of UML. I dreamed that UML would do something for poor people and the oppressed class. But it did not – they followed the capitalist road. After the 1996 Initiation of the People's War, I came in contact with the CPN (Maoist) and we discussed and interacted and I came to know the real way of communism. UML used to say they were for Marxism-Leninism-Maoism. But after the establishment of the multi-party system they didn't put this into practice and they actually dropped the title of Maoist from how they describe themselves.'

Surya then talks about how there was a big change after the Initiation:

> 'Before, the organizations were all legal and the work was mainly propaganda. Then the Initiation really gave hope and belief to the people. The teams of the People's Army and the Party exposed and attacked bad elements and threatened snitches [informers]. There was coercion as well as struggle. They struggled with spies to stop their bad ways. But if they did not stop, force became necessary. People were changed in this way.
>
> 'Now it is increasingly necessary to organize in an underground way and there is a need for constant security. After the Initiation there were mass arrests and killings in this area, and many people were forced to go underground. Now our organization has to be more systematic – we have to find ways to use individuals in the best way, according to their skills and capacity. Organizing of different mass organizations has continued – like the farmers' association, women's group, and youth

organization. The Party is now underground and the mass organizations have been forced to operate semi-underground.'

Before 1990 there was a one-party monarchical system of government in Nepal, the panchayat. Then in 1990, a mass anti-panchayat movement forced the government to institute a multi-party, parliamentary system. The king and his family retained a lot of power, including control over the Royal Nepal Army. But for the first time, there was the semblance of a less autocratic system. Surya goes on to explain:

> 'After 1990 the multi-party system was instituted and people thought they would now have a better life and opportunities. But this didn't happen and the gap between the haves and the have-nots only got bigger. There was a great crisis in the country, with Nepal being the second poorest nation in the world. This is one reason I was attracted to the revolution and saw the need for class struggle in order to achieve equality. Another reason is that I saw that all the political leaders in the government had become corrupted and did not represent the people. This situation made me dispirited, like a lot of other youth. I saw that these politicians have no love of the nation and they became servants of imperialism and Indian expansionism. As a lover of the nation there was no other way for me to go than to join the revolution. And what I have come to know is that the main source of corruption and repression and problems in society is because of the reactionary state power and system. And unless we get rid of it we cannot have any of our dreams come true.'

Before we finish this brief session Rohit says he wants to send a message of solidarity to people in the United States and let them know, 'Our movement is an international movement. We hope to become successful as one part of the world revolution.'

It's late by the time we finish talking and I'm glad to hear that Rohit's mother has prepared a meal for us. We go downstairs and, in the traditional Nepali way, we sit on the floor and eat with our hands. The food is very good and Rohit's mother keeps trying to get everyone to eat more. They tease me about my ignorance of Nepali eating customs, like the way to wash your hands after a meal. You're supposed to pour water over your hands, letting it spill onto your empty plate. But when someone hands me a vessel of water, I stick my fingers into it and everyone laughs. I won't make that mistake again.

A cultural squad from the People's Army will be holding a political and cultural program tonight and my hosts are trying to figure out how to arrange for me to meet and talk with them. For security reasons the squad will be arriving in the dark, right before the program, which will last two to three hours. And soon after this they will have to leave to travel two hours in the dark before they bed down for the night. This means we won't be able to meet with the squad until very late. But after months of planning and anticipation I hardly mind waiting a few more hours before my first face-to-face interviews with members of the People's Army.

After we finish eating it's time to go to the program. We step out into the darkness and take off on the path, single file. I need to use a flashlight, but somehow the others are able to climb the steep, rocky trail without any illumination. Effortlessly, they zip up the terrain. But this is all unfamiliar and new for me and I find myself having to concentrate, shining my light so I can see, with each step, where I'm putting my foot. In the complete darkness, my flashlight beam cuts a hole of light just big enough for me to travel through. And as fast as I can, I move through this small tent of light. Even though I can't see anything around me, my rapid breathing tells me we're going higher and higher up the mountain.

After a while we reach a plateau where people have started gathering for the program. With only two lanterns dimly lighting the area, it's hard to see what's going on. But I can make out the dark outlines of what looks to be about 100 people sitting on the ground. Other villagers are still coming up the hill. My translator, Shiva, and I sit down a short distance from where the villagers are gathered and within a few minutes someone gives us a mat to sit on.

We have to wait for quite a while. But I enjoy the cool, crisp night and take in the atmosphere of anticipation that wafts over from the growing assembly of villagers. It appears the squad hasn't fully arrived yet, although it is really too dark for me to distinguish the different figures moving about.

I'm squinting into the darkness, trying to get a better look at who's around me. Then suddenly, an image appears right in front of me, someone close enough to reach out and touch. It is my first glimpse of a member of the People's Army – in silhouette, the figure of a young woman in a uniform and cap, carrying a rifle on her shoulder. Soon I start to notice more figures who have entered the area, in uniform, armed with rifles. One is standing on the periphery, right behind us, guarding the area. Others are busy getting things ready for the program. People from the village are still coming up the steep path. And as they arrive, one guerrilla has been assigned to shine his light on the ground to help people with the last few steps to the top.

Villagers are still arriving when the program starts at about 9:30 pm. It is hard to estimate in the dark, but it looks like about 200 people are now gathered to hear and see this People's Army cultural squad.

Surya opens the program with a short introduction. Then he calls for a minute of silence for all the 'martyrs' – those who have been killed in the People's War – and everyone stands and bows their heads. One of the first speakers is a young woman whose husband was killed by the police. She is a leader in this area and explains the goals of the People's War and appeals especially to the women to join the revolution.

Next the People's Army platoon leader gives a speech about the importance of armed struggle. He argues that it is necessary to pick up a gun in order to defeat the enemy, but then says, 'The most important thing in defeating the enemy is the masses.' Interspersed between the speeches, the cultural squad performs songs and poems, accompanied by lively rhythms tapped out on a small, traditional Nepali drum. The first song they sing talks about how the 'blood of the martyrs must strengthen the people.' This is my first exposure to the revolutionary culture being developed as part of this revolution.

More people emphasize how villagers need to join and support the People's War. One tells the people, 'The People's Army will protect the people and the people must protect the People's Army.' There is also news of battles – of police and 'bad elements' being killed, but also of recent casualties sustained by the People's Army. One man tells the crowd, 'They are killing us in groups – but we are starting to kill them in groups too.' Shiva whispers to me that this is a reference to a recent incident in the western part of the country, where several policemen were killed by guerrillas. Other incidents are also cited in which the police have suffered defeat. Another speech encourages people to 'exercise new people's power' by taking things into their own hands, settling disputes, solving community problems, and administering justice.

The program is still going strong when we leave around 11:00 pm. The trip back down the mountain is a little easier, but I have to take a bit more care going down, so I won't stumble and fall. Again, I have to concentrate, shining my flashlight so I can see each step, and I try to keep up with the others who are practically running downhill with total ease in the dark. We get back to the house, go inside, and lie down. To my surprise, I fall asleep immediately.

Two hours later, a bit before 1:00 am, I awaken to someone's voice saying, 'Comrades, get up, they are here.'

I sit up and see that members of the People's Army squad are coming into the room. It is completely dark, except for two small candles – one set up on a table in front of me so I can write in my notebook. The squad members file in and stack their rifles against the wall. The room is very small and with about a dozen guerrillas, plus a couple of local people and Surya and Rohit, it's very crowded.

Some of the young women guerrillas come and sit right next to me on the bed. They are wearing People's Army uniforms – military green pants with many roomy pockets and matching jackets. The caps are sort of squared off on the top with a full brim – a big, bright red star declares from the front. Across from me sits the leader of the squad, a young man who I guess is about 25 years old. He looks very tired but welcomes me with a big smile.

For the next couple of hours members of this cultural squad tell me about themselves and their revolutionary passion. Almost all of them come from poor peasant families. The women are the first to speak and talk about the tremendous repression in their villages and how they came to join the People's Army. At first the women guerrillas seem shy and hesitant. But as each one takes her turn to speak I notice strength and determination in their calm manner. They have a character about them common to teenage girls around the world – the way they sit next to each other, whisper a secret, or fix one another's hair. But there is also a communal and disciplined way about them that comes from living and fighting together as a military unit. And I am struck by how seriously they are dedicated to this revolutionary cause.

As I hear their stories, I see how the poor conditions of peasant life, the way feudal society oppresses females, and heavy government repression have driven these women into the revolutionary ranks. The first young woman to speak tells me:

> 'When I was 16, 17 years old I thought, why are we so oppressed, economically and socially? I used to think, how can we solve all these problems in our families and in society? Around that time, in 1995, the CPN (Maoist) conducted a boycott of the elections and parliament and a cultural team of the Party came to our village. I came to know the way to solve all our problems and get free from repression. I got

involved in the cultural team and then joined the Party. At that time my parents had forbidden me to join the Party's cultural team, but I did anyway. In 1997 bad elements forced me into a situation where I had to go underground and now I am working in this cultural squad of the People's Army.'

Other squad members were also first attracted to the Maoists because of the Party's cultural work. And while some of the women had to rebel against their families to join the struggle, there are also stories of relatives who encouraged them to join the revolution. Almost everyone has a story about how the police had brutalized and arrested members of their family. One young woman whose father is underground tells me:

> 'In January of 1996 I was reading in class 9 and the police came to my village to arrest those who were doing a cultural program in our school. Our teachers were arrested and my father and my uncle had already joined the Party and had gone underground. Five hundred police raided our village and arrested just about everyone – even the children and old people. My mother was arrested and I was also arrested and kept in custody. There was so much repression by the police, so I joined the cultural team of the Party. And because of the exploitation and oppression of the poor masses, and especially that suffered by women, I was inspired to find a way to free the masses from such a situation. I found this was being done by the CPN (Maoist) so I joined the Party.'

The arrests, torture, and murder by the government are all aimed at getting people to reject the revolution in fear. But I am beginning to see how heavy repression, at least among some people, has had the opposite effect – making them even more determined to fight. One older man had been working with the Party for twelve years at the time of the Initiation. He was arrested and jailed for 26 months and when he got out he immediately joined the People's Army. Another young woman told me how her father is in jail and her uncle, aunt, and brother have all been arrested. She said, 'There was no other way except to take part in the People's War. So that's why I picked up the gun.'

The last woman to speak is a teenager sitting next to me on the bed. She starts off by saying that there was a lot of support for the People's War in her village and that her father has been underground since 1995. She then recounts how heavy police repression destroyed her village: 'The police came to our home and terrorized us. They raped women and arrested many people in the village. In 1997 there was an incident of great repression by the police and now in this village of about 26 houses, there is no one left. Everyone has been forced to leave and go underground.'

Some of the young guerrillas first got involved with the Party through the revolutionary student organizations. Many of them had to go underground after being arrested and targeted by the government. One young man says:

> 'I began getting into revolutionary politics as a student and became the district secretary of the revolutionary student union. I got involved with the People's Army cultural team with the understanding that this is the only way – through the People's War – to get rid of exploitation. In 1998 the reactionaries filed a case against me, charging me with treason. At the same time the Party provided me with a chance to join the Party and now I am with the People's Army. I am committed to this political line and believe that the multi-party, parliamentary system, as Lenin said, is the place where they show you the head of a goat and sell you the meat of a dog.'

This cultural squad travels in the Eastern Region, putting on cultural programs. But they are given military assignments as well. One of them explains,

> 'The members of the cultural squad also participate in armed actions against the reactionaries. We find out about and target bad elements like police or spies. And whenever the Party instructs us to do this, we survey the situation and figure out the problems we have to encounter in order to carry out the task. When there is no enemy present we can work openly. Sometimes the squad also works in the fields with the people. And sometimes we help to solve the problems of the people that come up in the village, for example, settling disputes, getting justice when someone has been wronged, etc.'

After each squad member has talked another leader in this area tells how she came to join the People's War. She says:

> 'I was involved in the student movement and got married in 1994. After the Initiation my husband got involved in the People's War and in May 1997 he was killed, leaving me and our three-year-old son. After I got married, I worked as the president of the women's revolutionary

association in the district. Last May I was arrested and put in jail. When my husband was killed I vowed to be a follower of his way and I swore I would pick up the gun that had fallen from his arms. Now the authorities are always searching for me and I have been forced to go underground.'

I remember that she is the one who, at the cultural program, had called on women to join the People's War, so I ask her why she feels so strongly about the role of women in the revolution. She answers:

'It is said in this society that women should work according to the wishes of their fathers, their husbands, and their sons. This is how society treats women. Capitalism exploits women and gives them no equal rights in property and in other aspects of society. This problem is not due to certain men or groups of people, but the root cause is the reactionary government, working with the expansionists and imperialists. It is clear we cannot get success in our struggle, solve our problems, and get rid of all kinds of exploitation and oppression as long as this reactionary government and system exists. We can only overthrow it by using guns, and this is why we have to wage People's War. Then we can have new forms of people's power where women can get equal rights.'

The squad has to leave soon so one young man says some final words: 'On behalf of the Party and the people involved in fighting the People's War, I give you our heartfelt thanks for coming from such a long distance to learn about our struggle and hope the message of our struggle will become known to people all over the world.'

It is now 3:00 am and the squad still has a two-hour walk to get to another village before sunrise. Before they leave, the squad leader says they would like to present me with a gift. The guerrillas get up, grab their rifles from where they have been leaning against the wall, and line up in the small area of the room next to the beds. Someone gives a command and they snap to attention, holding their rifles to their sides, eyes front. The leader of the squad steps forward and I jump off the bed and take two steps toward him. He extends his arms and I see that he is holding a khukuri, the traditional, razor-sharp, curved-bladed knife of the peasants in Nepal. It is now being used by the guerrillas to fight their enemies. He hands it to me and says, 'We would like to present this to you; it is our symbol of war.' We will not be able to travel with this weapon so someone takes it from me, with the promise that it will make it back to Kathmandu where I can retrieve it.

The squad lines up to say goodbye and each one takes their turn and steps forward to give me a *lal salaam* – the revolutionary red salute. First they raise their right hand in a strong fist, then extend both fists down in front of them, then reach out with both hands to give me a handshake. In a very serious manner, they take my two hands in their two hands in a solid grasp. By any measure, this revolution is up against tremendous odds and each of these young fighters is risking their life every day. But somehow they are fearless and confident in the justness of their cause and that it will be victorious.

The squad files out of the room very quickly and I can't even hear them as they go outside and quietly leave the village. All of a sudden, the room is empty, now darker with only one small flame near our beds. I blink a few times, peering at the empty space which, only minutes earlier, had been crowded with green uniforms. I tell myself yes, it really did happen, you spent the last few hours talking with guerrillas from the People's Army.

It is a good bit after 3:00 am now and we will have to leave the village early in the morning, so we blow out the last flickering flame, lie down, and try to get some sleep.

* * *

JOURNAL ENTRY

March 21, Sunday

Yesterday a front-page article in the *Kathmandu Post* carried the headline, '7 Maoists killed in encounter.' According to the brief article, the incident happened in Banepa. I read the following, keeping in mind that such news reports about the People's War in the mainstream, pro-government press may or may not be reliable: 'The Maoists were burned to death as the bomb they hurled at the police exploded among themselves after striking against the wall of the house they were staying in. The Maoists reportedly threw the bomb after they were told to surrender by the police. According to police, the cross-firing continued for two hours ...'

Since I've been in Nepal, there have been many reports of guerrillas being killed by the police, as well as of police or others being killed or injured by Maoists. Each time I read these items I am interested in knowing where they happen. So on this morning I get my map out to locate Banepa and find that, in fact, it is right along the road we had traveled on to the Eastern Region. I don't think about this too much more during the day – although the news continues to occupy a corner of my mind.

In the evening a friend comes by who had been with us on our trip to the east. He had been back there since and talked with people about the Banepa incident. He tells us that the people killed in the incident were from the cultural squad we met.

I am stunned by the news and immediately close my eyes and try to picture their faces. Late, in the middle of the night, we had crowded onto the bed together, our legs crossed, knees touching. In the candle-lit room, their shadows had loomed large on the walls. I concentrate on remembering the faces of the young guerrillas who had left their villages after seeing their families and friends arrested, beaten, raped by the police; their fathers, uncles, mothers forced to go underground before them. Only a short time ago they had shared their war stories with me. Now seven of them, four men and three women, were dead after refusing to surrender to the police.

March 25, Thursday

Today I learned a little bit more about the incident in Banepa. The *Kathmandu Post* had reported they had died after they threw a bomb at the police. But people from the Party tell me they were surrounded and refused to surrender, and there was shooting back and forth. Then the police set the house on fire and when the guerrillas were finally forced to run out they were shot. According to the Party, the murder of these guerrillas was due to the spying and informing of an UML person who is running for office in the elections. They say this is typical of the role the UML is playing these days, directly helping the government to target and murder Maoist revolutionaries.

This afternoon someone brought me photos of the guerrillas killed at Banepa. One of them was the young woman with the baby face who had been sitting next to me on the bed that night. Her eyes had shone brightly, even in the dim flicker of candlelight. I had heard my first notes of a revolutionary song in Nepal from these youth. They had been the first to teach me their *lal salaam*. They had given me my first face-to-face conversation with members of the People's Army. They had lived such short lives. But I am told that, like other martyrs in this revolution, they will be remembered by all the people who continue to wage this guerrilla war.

I will certainly never forget them.

2
Villages of Resistance

In the countryside of Nepal, it's hard to sleep past 5:30 or 6:00 am. Roosters start crowing and the rustle of those already up invades your sleep. In that world of semi-consciousness before waking, I hear the women stoking the new day's fire, the men getting ready to face the fields. Someone brings me a cup of hot milk-tea and I sit on the edge of the bed sipping it and thinking about how only three hours earlier, members of the People's Army had been sitting here next to me, telling me about their lives.

About 7:00 am, several women appear at the doorway and we motion them to come in. They are from the village and want to talk with me.

From the time they were very young, these women have worked in the fields, hauled water, and cut grass to feed animals. All of them talk with bitterness of how girls are usually not allowed to go to school. All of them speak with optimism of how the People's War is giving them new opportunities – not only to get an education but to join a revolution which they are convinced is the way to a better life. One woman says:

> 'We are illiterate. Due to our traditional customs, we did not learn to read and write, because it is said daughters should not be educated. But now we are beginning a new people's education. Before we only passed the time working in the fields, bringing fodder and grass to the cattle, and doing other household work. The main thing we have come to know is that all the oppression we are facing today is due to the reactionary state power. We have come to this conclusion and what we are doing here has made us clearer about the exploitative nature of the reactionary government.'

The women tell me that according to feudal tradition, parents arrange most marriages – young people have no say in who will be their husband or wife. In general, traditional customs discriminate against women. But now, with the spread of the People's War, the women say things are beginning to change. One woman, who looks to be in her early twenties, explains how new ideas are beginning to take root:

> 'There is struggle in the family because sometimes there is uneven understanding. If the parents understand things, it is easy. But if not, it is difficult for a woman to take part in the People's War. But today in this village, everyone is taking part. There are only two or three houses that are not participating. We became active in the revolution with the conclusion that everyone should be equal and united for the welfare of the people. Husbands are also now beginning to do jobs they never did before like cooking, washing dishes and taking care of children.'

I ask the women if they will let me photograph them. At first they seem a little shy at the idea, but then they smile and say yes.

As I take their pictures I can't help thinking how these women seem to have been *changed* by the hope this revolution is giving them.[1] I have been in Nepal long enough to see many other poor women, first in the city, then as we traveled through the countryside. My camera had captured images of vendors squatting beside piles of vegetables or tourist trinkets. I had given rupees to women begging in the streets with their kids. In the countryside, I crossed paths with peasant women hauling heavy loads up mountain trails. I saw the tired faces and discouraged eyes of women stoking fires as the sun began its next ascent.

The women in this village are no doubt just as poor and experiencing as much, perhaps even more, hardship and poverty than the people I've seen outside this guerrilla zone. But the faces of these women have a very different expression. Here, the women seem confident, their body movements less heavy. They look happy and, at the same time, deadly serious. They have a vision of a better future that comes through in the edge to their words. Their hatred of the police and others who oppress them informs the hard look in their eyes. Their faith in this revolution gives them a physical confidence in how they move about.

We arrive at another village late in the day and are taken to a house nestling on the mountain between dense forests and terraced farm fields. We go inside to a second-floor room that's catching the day's last rays of sun. Several people have already gathered and Dipak, a young supporter of the People's War, begins telling me about this village. He is only 18 years old and, unlike almost all of the people I will meet in the countryside, speaks very good English. He informs us that he's studying to be a doctor so he can serve the people in the countryside. The first thing Dipak talks about are some of the people who have been killed in this area.

'Rewati Sapkota was 25 years old, married with a son and a daughter, at the time of his death in 1998. He had been in the People's Army a year and a half before being killed by a police commander in a sudden encounter in the forest. Four guerrillas were attacked by about a dozen police and Rewati and one other squad member were killed.

'Bhim Prasad Sharma was 20 years old and had just graduated from high school and gotten married when he was killed – in another sudden encounter with the police. In this incident, eleven guerrillas were attacked by more than 100 police and all but Sharma were able to escape. He had been in the People's Army only three months at the time of his death.

'Sabita Sapkota was 21 years old when she became a martyr and had just graduated from high school. She had to rebel against her family to join the People's Army and after she went underground she didn't see her family again because they would not support what she was doing.

'Binda Sharma, a 25-year-old woman, killed in 1998, also had to rebel to join the People's Army. Her husband didn't, and still doesn't, support the People's War and in fact now works as a police detective in Kathmandu. For over six years, Sharma had been in this arranged marriage. But then one day, after she began working with the local Party, she ran off and joined the People's Army.'

Dipak then talks about how the People's War has been developing in this area:

'There are 60 to 70 families in this village and about 80 percent here support the People's War. Two weeks before the Initiation a policeman was killed and then the police arrested 70 people. The whole village was raided by more than 150 special force police. One policeman was killed and the day after, this house we're sitting in was raided. The police fired six times into the air and the local people started to rebel. The police attacked many people with sticks and two people in the house were arrested. Five days later they raided the whole village. They went house-to-house at 1:00 am. Children only nine or ten years old, up to older people over 70 years old, were arrested. The police would come in and grab people and take them to the local police post. Male police were grabbing at the women. In all, about 60 to 70 people were taken into custody and all of them were charged with killing the one policeman. Actually, at the time, this cop was only "missing" – only later was he found dead. This happened over three years ago and 15 people from this incident are still in jail, without trial.

'All this really affected people in the village and made them very supportive of armed struggle against the police. So when the People's War started there was a lot of activity here – including wall postering, distribution of the communiqué from the Party, and torchlight processions. Fifteen to 20 people went underground and some were caught by the police in the first week. Now there are about 15 people from this village who are still underground, including my father.

'In the three years of the People's War there has been continuous revolutionary activity in this village and things have moved on. There have been actions against bad elements, many processions, and a lot of postering and wall paintings. The Party has held condolence meetings for the martyrs and many mass meetings to explain the goals of the People's War and to build support for the revolution. People have joined the Party and the People's Army and, in response, the police have carried out many raids and arrested many people. This house has been raided five times.'

More and more villagers have been coming into the room and the floor space is now packed. Others are crowded around the doorway, trying to get a peek and listen in. We decide to take a dinner break and then reconvene in a larger room upstairs. The women in this house have ready a big meal of *dal bhat*, fish stew, and curry potatoes. They have even prepared a special dish for me – French fries, which people here have never eaten, but have heard is a favorite dish in America. We eat quickly – by now quite a few people have arrived from neighboring villages and are waiting to meet me. After we finish eating everyone goes upstairs into what looks to be a room for storing grains. We sit on mats that have been set up next to stacks of harvested corn. I look around at the gathering of about 50 people and notice that it's mostly women, along with their kids (who seem younger than twelve), and a few old men. A woman gets up and starts by saying:

'We would like to thank you for coming and we express our solidarity. We face many kinds of repression. The reason why it is mainly women here at this meeting is because all of the men are underground so they could not come tonight. And because of the police we could not have this meeting in the daytime. We have been raped and beaten in custody. We have no way but to fight back with the Party, which is fighting against this oppression. We women have to work on our own small plots of land and because of the repression from the reactionaries we have a very hard life.

'They have killed a small child of only five years old and an old man of 90. They have raped a ten-year-old girl and a 70-year-old woman. They have looted our property. They have taken the property of even the old people. They have raided homes and put people in jail. There are so many cases of those who are fighting for justice being targeted by the police. In the elections they force us to go out and vote, even though we don't want to. There are massive violations of human rights. Men who speak out, demanding the government act according to the law and constitution, are being hunted down. And women are also being forced to go underground. Sometimes the men are separated from their wives for one or two months and then the police come and interrogate and rape the wife.

'Because of this terrible situation created by the reactionary government we have come to know, and it is our compulsion to understand, the need to pick up arms and fight them and wage a successful People's War and build new forms of people's power. We can't be free from this inhumane treatment and repression until we get rid of this reactionary government. And this is the reason we support and have joined the People's War. One way we support the People's War is by collecting crops and money, according to people's capacity, and sending it to those who are underground, who are away from our village.

'Even though they are killing our people, at the same time, the People's Army is fighting and we are hopeful that they will increase their capacity to fight even more. We really believe that the murder of our comrades by the police cannot stop the People's War. The blood of the martyrs is the fuel of the revolution. Understanding this, we are united and our unity is sure to defeat the enemy. We strongly believe the People's War will be victorious. On behalf of the women's association and the farmers' association, we would like to say welcome and thank you for coming so far to learn about our lives.'

Other villagers also get up to speak, but the meeting has to end after only about an hour because people have a long walk back to their houses and, for security reasons, must travel without any light. They pledge to keep the secrecy of the meeting so the local police or other people unfriendly to the revolution won't find out about it. And then people line up to shake my hand. Some young girls, who look to be about nine or ten years old, purposefully elbow their way past the adults. Instead of giving me the more traditional Nepali *namaste* greeting, which is with your hands held together as in prayer, they hold their fist up in a *lal salaam*, the red salute, and then clasp my hands very tightly with their little hands.

3
The Raid on Bethan

Nepal is a country of remoteness. Most of the countryside is accessible by just one form of transportation – walking. The whole country has only about 8,000 miles of roads; these include some 3,000 miles of routes inadequate for vehicles. There are no river navigation facilities and only one, 30-mile railway linking Janakpur, in the eastern Terai area, with Jayanagar, just over the border in India. In the countryside, whenever I ask someone how far away we are from another village, they never answer me in distance. They always reply in hours or days – which means how long it takes to *walk* there.

There are just two main highways in Nepal: the Mahendra Highway, which runs the length of the country from the Indian border at Kakarbbhitta in the east to the Indian border at Mahendranagr in the west; and the Prithvi Highway, which links Kathmandu with Pokhara, which lies to the west of the capital. Buses crammed with people constantly crisscross these roadways. People also travel on buses on the limited number of roads which branch off into the countryside.

This extreme lack of infrastructure reflects once again the profound poverty of Nepal. And for the people in the countryside, this lack of good transportation is not simply a problem of 'inconvenience.' More importantly, it is a question of *safety*. The buses, which are almost all made in India, are old and worn out. And when you add this to inadequate roads, steep terrain, and overcrowding, the result is frequent and horrendous accidents and many fatalities.

At the same time, the remote conditions of the country are very favorable for fighting a guerrilla war. For the lack of roads into the countryside makes it hard for the government to bring in armed forces in large numbers to fight the guerrillas.

We get on a bus in the Eastern Region half an hour before it's scheduled to leave and most of the passengers have already boarded – at least those who got here early enough to get a seat. The overhead bins are stuffed with all kinds of packages and the aisles are packed with sacks of grain and big containers of what looks like kerosene or some other kind of liquid fuel or oil. Luckily, I am able to get a seat. Otherwise I would have ended up sitting for hours, teetering atop a bag of millet or a drum of fuel. More people keep piling in until the very last minute.

There are all kinds of people on the bus, but many more men than women. Some are neatly dressed and look as if they are returning from a visit to the city or are perhaps now transplanted city dwellers coming back for a family visit. Many of the young men are wearing jeans and T-shirts, while others sport more traditional Nepali loose-fitting pants and tunics. The extreme poverty of one young boy, who seems to be by himself, is revealed by his tattered and dirty clothes.

By the time the bus pulls out, bodies are crammed in tight. A woman nursing her baby, and clutching another small child, balances herself on the sacks of grain piled next to me in the aisle. Like everyone else on the bus, we are pressed up against each other and she's almost sitting in my lap. The bus takes off pretty much on schedule and I see the road we'll be on until way past dark – very dusty and bumpy, winding up a steep mountain. There are very few and always short straight-aways and lots of hairpin bends. The bus lurches up the mountainside in fits and starts.

We haven't been going too long when the bus comes to a sudden halt. There is a lot of conversation and then several men jump off the bus. There are maybe a dozen people also riding on top of the bus and some of them jump off also. This is to be the routine for the next several hours: The bus gets to an incline that's too steep for its weight and so a bunch of people have to jump off, walk up the road and get back on after the bus has made it up a steep part of the road with its somewhat lighter load. Sometimes it takes several attempts before the bus conquers a particularly sharp incline, with successive appeals for more people to get off the bus, until it is light enough to go forward. I'm already starting to wonder if this constant delay has been calculated into the scheduled 'three hours' to our destination, but from the calm attitude of people around me, I take it that this routine is not out of the ordinary.

There is a crew of about three young guys who work with the driver to navigate the bus up the winding road. I try not to think about it too much, but the road is very narrow and in many instances, when I look out the window, I can see that the bus is right on the edge of a very sharp drop down the mountain. One miscalculation by the driver or miscommunication from the guides could send us careening down this mountain. The guides are constantly jumping on and off the moving bus – for most of the time the bus is only going about 10–20 miles an hour – and they are covered from head to toe with the light reddish-brown dirt from the road. Soon much of everything in the bus is also covered with many layers of this dust.

The guides have a system of signals they give to the driver, usually by banging on the side of the bus – one, two, or three bangs. When they are on the bus, hanging out the door, they have a button they press which sounds a horn. In this way, they tell the driver when the bus is getting too near the edge, when it has to stop and back up in order to make a turn or make other maneuvers to get the bus steadily up the mountain. Sometimes the bus has to back up for quite a way, after coming to a halt, in order to gather speed and 'make a run' at the next steep part of the road. After about an hour and a half of this, I can see that we haven't really gone that far – and I know our destination is quite far up this big mountain. After about two hours the bus breaks down completely and the crew starts pulling out wrenches and other tools to make repairs. Meanwhile another bus comes down the hill. I can't believe it is going to pass us on this narrow road, but it does. After a 45-minute delay for the broken part to get fixed, we resume our stop-and-go lurch up the mountain. In this way, our three-hour bus ride takes over six hours.

Some people are waiting for us when we finally reach our destination and we immediately head off along a rocky path. We are high up in the mountains and the sky is bursting with stars, but it's very dark. After about 45 minutes of walking, even though it is too dark to see very clearly, I can tell we are in a village, winding in and out of yards, stepping over piles of hay and encountering oxen, bedded down for the night. We reach the house where we'll be staying and the people here immediately prepare a meal for us. By the time we finish eating it's late, but our hosts want to have a little discussion before going to sleep. They are excited we are here and can't wait until morning to talk.

This area in the Eastern Region is where some of the most advanced actions took place at the start of the war and the Party's work here continues to be very strong. One of our hosts explains that this village, on the boundary of a large area of about 150,000 people, is developing towards becoming a 'base area.' He says that within five to seven hours, they can mobilize a mass meeting of 5,000 people. In such instances, he says, the local police will not dare come in without reinforcements. And by the time they arrive, the meeting will be over and people will have already dispersed.

We stay in this village for several days, but because of security concerns, we're confined to the house during the day. Several people come to talk with me and I know arranging such visits is very difficult. Many people, because they are wanted by the police, can only travel at night, and even then they are risking their lives. Communication is hard and many different factors have to be taken into account – the security of the whole area, of the family we are staying with, of the people who come to talk, and how long it is safe to stay in one place without calling attention to ourselves. We even have to be careful when making quick trips to the outside latrine. The People's War is strong here. But this means government repression is intense. The house we are staying in has been raided many, many times by the police.

Compared to Kathmandu, life is very different for the 90 percent of the people in Nepal who live in the countryside. Daily life here is hard and routine. It starts very early, around 5:30 am, when the women get up to make tea for the household. The adults go to the fields or tend to other chores. Around 10:00 am they come back for a meal, about the time the children go to school. After this, people go back out to work for the whole afternoon and return when the sun is low on the horizon. The evening meal is usually late, around 8 o'clock, and by 9 everyone is getting ready to go to sleep.

The house we're staying in is a three-story clay structure, with dirt floors, simple wooden stairs, and shuttered windows. There are many people living in these five rooms. There is the mother and father of the house and their children – several grownup sons and daughters. And then there are the sons' families, each with several small children. The father's two sisters also live in the house. One is a widow, the other left her husband because he beat her.

I figure there are about 16 people living in this house, which means two or three people must sleep together on each bed. But I am given a bed of my own at night. And there are many other ways in which the family is hospitable and generous. They are poor, but willing to share everything they have – constantly asking us what we want to eat, even though it's hard for them to feed their own family.

Cooking is done on the first floor of the house. In one corner a fire is built and mats are placed for people to sit on and eat. In this home feudal tradition seems to still govern mealtime. Usually the men will eat first, then the women and children. The 'mother of the house' sits and serves people the whole time. In another corner of this first floor, the family chickens have their domain. Then on the second and third stories are the small sleeping rooms, about 10 feet square with enough space for maybe two beds with a few feet in between. Everything is simple and basic, but very neat and clean.

The family we are staying with goes about their daily routine, but along with their comings and goings they poke their heads into the room to see if we need anything – and to catch a bit of our discussions. Everybody is eager to talk and listen and it strikes me how *engaged* everyone is in the People's War. Even the little kids hang around the fringes of the conversation to listen. The women are shier, but eventually, all of them come by and introduce themselves.

Right after the war started, the police raided, arrested, and tortured people in the Kavre district and killed a number of people. One of them, Tirtha Gautam, is known throughout the country and promoted as a brave hero.

In the afternoon, the brother of this famous martyr comes to talk with me. He tells me that Tirtha was only 33 years old when he was killed. He was a leader in the Party and the People's Army. He was a member of one of the Party's Sub-Regional Bureaus in this Eastern Region, the Secretary of the Kavre-Ramechhap District Organizing Committee and the Military Commander of the district. He had either led or been associated with all the leading guerrilla actions in the Sub-Region since the time of preparation and Initiation of the People's War. Born into a lower-middle-class peasant family, Tirtha Gautam had worked as a schoolteacher before becoming a full-time revolutionary in 1988. At the time of his death, he had been active as a revolutionary communist for more than a decade.

Tirtha Gautam was killed while commanding an attack on a police outpost in Bethan, in one of the most backward hilly regions, some 60 miles east of Kathmandu. This was the first successful raid on a police post after the Initiation of the People's War.

This is the story told by the Party: On January 3, 1997, in the dark of night, Tirtha led his squad of 29 guerrillas into battle. Equipped with homemade guns and bombs, the guerrillas surrounded the police post and ordered those inside to surrender. The police shut themselves in the building and started firing, setting off a pitched battle that lasted for several hours. Tirtha Gautam was shot in the head and died instantly. But this inspired his squad to mount their attack more vigorously and they succeeded in overpowering the police. Two policemen were killed and two others seriously injured. The guerrillas seized four rifles and hundreds of rounds of ammunition. In addition to Tirtha Gautam, two other squad members, Dilmaya Yonjan, a young woman, and Fateh Bahadur Slami, also lost their lives. The squad carried them away and returned to safety. The police mounted a vicious counterattack – search helicopters hovered over the remote mountain forests and the police carried out a combing operation in the region searching for guerrillas.

The widow of Tirtha Gautam, 30-year-old Beli Gautam, comes to see me with her two young sons, ten-year-old Delip and eight-year-old Tanka.

Beli Gautam has a hard time speaking and says she is having trouble with her throat. I notice right away that she is rather shy and quiet. But Delip speaks up with a confidence and conviction beyond his age. He sits up straight, looks us in the eye, and says with force, 'Our mother says we must follow the path of our father and when we get older, we will have to go fight.' Beli tells me she also has two daughters, a five-year-old and a twelve-year-old, and that while the younger one is too little to understand much, the older one already wants to fight in the People's War.

The police have raided the house of this family many times – sometimes in the middle of the night – saying they are looking for guns and bombs. Once, soon after Tirtha was killed, the police came and asked the family, 'Now that we have killed Tirtha, what will you do?' They seem to have been hoping the family would feel threatened and say they no longer supported the Maoists. But young Tanka said to them, 'You killed my father, now I will kill you.' The police replied in a mocking way, 'Oh, and how will you do this?' They handed Tanka their rifle and said, 'Do you know how to use it?' To their surprise, Tanka took the rifle and began to cock it, showing the police that in fact, he did know how to use it.

Delip tells me of another time when the police came to the house and started busting everything up. When they found a guitar they asked, 'Who plays this?' and then destroyed it. He also recounts the time they threatened to arrest his grandfather. His grandmother stepped forward and defied the police, saying, 'If you are so brave why don't you kill him now with your gun.' And then a police officer stuck his stick into the grandmother's mouth to shut her up.

Before the family leaves, Tanka sings two songs for us, the first he wrote himself. His young voice is sweet, but earnest, and the song has a somewhat melancholy but spirited tune. It tells of how his father was killed in the raid on the Bethan police post and how, before he left on this mission, Tirtha swore he'd bring the ashes of the police post back to the Party. The second song is sung to the tune of a Nepali folk song, but Tanka has given it new lyrics. He sings about the police and the repression the people face and then declares that the people's answer to all this is, 'Give us the gun of Gautam,' referring to his martyred father.

After Tanka finishes singing, Delip says, 'When the police attacked the squad my father was so brave. Now when the people see a fire, they remember the heroic raid on the Bethan police post. When they hear a loud sound, like thunder, they remember the guerrilla's guns in Bethan. So every day, when the people see these kinds of signs, they remember this incident and my father who sacrificed his life.'

It's now time for the two boys to go to school and as they head for the door, they turn around and give me a red salute, shooting their little fists into the air.

4
Rifles and a Vision

When the People's War started two categories of armed groups were organized – fighting squads and defense squads. In the beginning, these fighters were inexperienced and armed with only some single-shot musket rifles and khukuries. But in the course of three years of war, a qualitative development in army building had been achieved. Now, in 1999, the military forces of the People's War are classified into three categories: main force, secondary force, and basic force. The people's militias, operating in the local areas, are the basic forces; guerrilla squads with about nine members are the secondary forces; and the main force at this stage is the platoon, comprised of about 27 guerrillas. Efforts are well underway to develop these platoons to the level of companies made up of about 100 fighters.[1] A platoon in the Eastern Region will be holding a mass meeting near the village where we are staying. They are scheduled to arrive late tonight and Shiva and I will get a chance to talk with them.

A couple of hours after our evening meal someone comes to tell us it's time to go and we set off, in single file. Soon, I see a crowd of people on the right side of a house in an area illuminated by a single, bright lantern. When we get close, someone motions for Shiva and me to go around the house, to the left. Everyone else peels off to directly join the gathered group. We are moving at a very fast pace, not quite running, but my heart is pounding, not knowing quite what to expect. We come round the side of the house and right before us is a platoon of the People's Army, their backs to us, lined up in formation.

As soon as we appear a voice rings out a command and the guerrillas snap to attention, their rifles raised to their shoulders. The platoon is standing in two lines and we walk down the aisle they have formed up to where several people are waiting for us underneath a bright lantern. When I reach the front, some army and Party leaders greet me, one at a time, putting garlands of flowers round my neck and handing me presents. Each one grasps my hands firmly and gives me a *namaste* and a *lal salaam*. A young woman, a leader in this district, draws me quickly to her and gives me a strong embrace. The platoon commander also gives me a big hug. All the while, a platoon member darts around snapping pictures with a small flash camera.

We go inside the house, take off our shoes, and enter a small bedroom that has been prepared for the meeting. A table has been set up in between two beds, decorated with a bright cloth and two vases of flowers. Shiva and I sit on one side of the room on the bed and everyone else piles onto the other bed. Most of the platoon members remain outside, but two of them follow us into the room, a young man and young woman, who stand to my left, guarding the door with rifles up on their shoulders. The Party's district leader comes in and sits to our right. The platoon commander takes his revolver out of his waistband, places it on the table in front of us, and sits down next to her.

There are formal introductions and I find out that most of the ten or so people scrunched together on the bed opposite us are local leaders. The two guerrillas at the door are also introduced and they each give me a handshake. Everyone stands for a minute's silence for those killed in the revolution and then three platoon members come in to sing us a revolutionary song. Shiva whispers the words of the song in my ear:

There is no certainty in life
We may die today or tomorrow
We may depart and hope to meet again
There are so many martyrs
We are proletarians and have rifles and a great vision
And we do not like to leave each other
In war there is bloodshed and insecurity
We tell the fathers and mothers
Don't weep, there is the sound of a bullet
and people are falling and there are martyrs
We are internationalists
Fighting to free the whole world
History will be victorious for the people
To those who die, following the path,
To those who die, fighting the enemy.

Since we have a limited amount of time I focus my questions on three things: 1) what it was like here at the time of the Initiation; 2) government repression; and 3) current efforts to establish base areas.

The district leader tells me that before the Initiation the Party was already strong here and they were able to carry out the plan of initiating armed struggle with much success. She explains:

> 'The Party made preparations to launch the armed struggle and at the same time we were doing open mass work and making people aware of the political situation. There was a lot of good political work done among the cadres about the need to pick up the gun and this helped us to make the leap from mass struggle to armed struggle. When the Party came up with the plan to go to armed struggle we targeted the symbols of the enemy – the police, feudalistic landlords, compradors, and imperialism .... There was a raid on a police post and sabotage of landlords who had exploited the people. We completed the First Plan of starting armed struggle and the Party gave us the next plan. This Second Plan was to wage guerrilla warfare, so our plan was to seize weapons. We successfully raided the police force in Bethan and in that action two police were killed and we seized 200 bullets and some rifles. Three brave comrades were killed, including Tirtha Gautam, who was a district committee secretary and the commander of a squad.'

For the last couple of months the CPN (Maoist) has been implementing the Fourth Strategic Plan of developing selected areas toward becoming base areas. The district leader talks about how they are carrying out this plan in this area:

> 'Now in the Fourth Plan, we raided a police post in this district in which one policeman was killed and we were able to seize four rifles, one made in the USA. We were also able to raid a police post where three policemen were killed and four rifles were seized, with no harm to our side. This action was done in the memory of the martyrs in Bethan. In all, we have lost some squad members and political leaders – 25 martyrs in all. But at the same time we have been able to take out 14 of the enemy – spies and bad elements, six who were police.
>
> 'We have designated certain areas as proposed base areas. The first thing we have to do in these areas is wipe out the enemy. If there are police posts we try to remove them, raiding them, taking guns and running the police out. The second thing is if there are spies and informers we will remove them also, which can mean elimination. And the third thing is to remove the feudal landlords. In the main zones we will do all these things. The other thing is we have asked all the leaders of the government Village Development Committees [VDCs set up by

the government],[2] who have been elected, to resign. And we have appealed to all the masses to boycott the elections. We have done all this with support from about 75 percent of the people. We went to the VDC chairmen and asked them to take part in the new forms of people's power, and none of them has refused to resign. We are centralizing and concentrating our forces to constitute and manage new forms of power. In these areas we are making different forms of mass organizations – among women, farmers, students, and children – in order to organize everyone.'

Next I talk to two groups from the platoon. Most of them are young, in their late teens or twenties, but there are a few older guerrillas, even two or three in their forties. Just about all of them are from very poor peasant families and have experienced or witnessed a lot of police repression and brutality. Several first got involved in the revolutionary struggle as students. Two young men are from an oppressed ethnic nationality and they talk about how the suppression of their culture and language was a big factor in their joining the revolution. Many of the guerrillas reiterate the point that they took up arms because they saw no other alternative. A 34-year-old man explains what had compelled him to join the People's Army:

'I became interested in the Party and the People's Army when I came to analyze the situation in our country – that there is no health care, jobs, etc. Even the youth who leave the countryside and go to the city can't find jobs. And I asked why does this situation exist? I came to see that it's because of the reactionary state power that is exploiting people in many different ways. And I came to see that there should be class struggle and that the poor will be able to win out because our numbers are big and we should be able to overcome all difficulties in organizing the masses of poor people.

'Then I met the Party, it started the People's War and I joined the People's Army. The second reason I came to join the People's War is that the reactionary government couldn't give us a chance to have a peaceful movement. When people went to them with their demands the government repressed the people and didn't even follow their own laws. In this way the reactionary government gave us no way but to start the armed struggle.'

There are only a few women in this group and they speak next about how they got involved and why they were attracted to the People's War. An 18-year-old says:

'I come from a very simple family with a small plot of land. I feel strongly that there is no other option, if we want to wipe out inequality, the great oppression of women, expansionist India taking over our land. All this cannot be solved unless we fight the reactionaries. So in my heart I came to the conclusion that I had to pick up the gun. The conditions in society are such that there is inequality between men and women. Sons can get the property of their fathers, but daughters cannot. If we get married and go to live in our husband's household there is suppression, so we can't be free. As far as the People's War and the Party are concerned – they are working for equality between men and women. Women are not denied any aspect of the revolution, and I'm convinced that the new democratic system we're fighting for will provide equality for women.'

Several guerrillas say they have been inspired and emboldened by the bravery of those who died fighting. One 40-year-old platoon member says:

'I am from a poor farmer's family in Bethan. I did not get a chance to get an education, so I went to the city to work. When I was in Kathmandu I came to know about communism and I talked with many revolutionaries and became convinced that I should become a communist. So I got involved in the student movement, met some of the leaders and then joined the Party's youth organization. I joined the People's Army on the day of the raid on the Bethan police post. The main reason I joined was not only economic repression, but being from an indigenous ethnic group. We can't speak our language, read our mother tongue and are repressed by the Hindu government. So now I have great hope and determination that we will be able to establish a new democratic system that is for equality and will wipe out all the discrimination being done by reactionaries.'

An 18-year-old woman recalls the impact of the raid at Bethan on her decision to join the war:

'I was at home and in the revolutionary student movement when the People's War was initiated. I was in 8th grade and was working as a member of a people's militia. I was very close to one of the martyrs killed in the Bethan raid. She was a friend in the revolutionary student movement, and when the People's War started she felt she had to take up more responsibility. When my friend went to go raid the Bethan police post she thought that if she became a martyr it would be a sacrifice to the women's movement and that this was her duty.'

It is already way past 3:00 am and we have to end. The district leader and platoon commander clasp my hands and both give me a tight hug. It feels as if through this brief physical contact, they are trying to convey the life-and-death nature of their struggle and their desire to make a connection with people outside of Nepal. They leave quickly and quietly.

After cups of tea are passed around we wait a bit and then we too head out into the darkness. At this hour we can't use our flashlights and have to walk quickly in the dark. One of our hosts offers me his arm and guides my steps. We are going quite fast and I have to surrender myself completely to his skill, my feet moving swiftly over terrain that surprises me with each step. I stumble a few times and walk into some branches along the narrow path, but my guide seems to know exactly how to show me the way with the slightest movement of his arm and I am surprised how I can move so fast without seeing at all where I am going.

When we arrive at our next shelter, an older man lets us in and motions us to lie down on two beds which have been vacated for us. Shiva falls asleep immediately – it's now past 4:30 am – but I can't get to sleep for quite a while.

A few hours later, we get up and I thank the father of the house for giving us a place to sleep. He tells me, 'No, it is nothing ... whatever I can do to help the struggle.' The family prepares a meal for us and then soon after this, someone comes to take us up the hill to where we will catch the bus. As we start out I look around and for the first time see the beauty of the terrain I had traveled in the dark. We start off weaving through the small village, going past women sitting outside working, piles of wheat lying outside to dry and oxen and goats lying down munching grass. Then we head out into the vast fields of grain, where the paths are very narrow – there is barely just enough room to walk along the rows of precious food.

I look out in the distance across the green, terraced landscape and see a large group of people who are working on the land. It's the platoon we met last night and they are clearing an area to build a 'memorial to revolutionary martyrs.'[3]

Incredibly, just as we reach the road at the top of the hill, the bus pulls up and we clamber on. There is only a brief moment in which I get a chance to look back and wave goodbye and then we are off.

5
Revolutionary Work in the City

Living with people in the countryside and seeing the daily life of the peasants is giving me a vivid understanding of the severely lopsided state of the world. Rural life here is so primitive, the people must work so hard, and still they remain so poor. Only 10 percent of the country has electricity. Conditions of life, even in the city, and even for more middle-class people in Nepal, are so difficult compared to the US, where there is so much wealth and technology and a huge section of middle-class people live in such relatively luxurious conditions. Even poor people in the US commonly have so many things that 90 percent of the Nepalese people, who live in the countryside, would only *see* if they went into the city – cars, TVs, radios, refrigerators, running water, electricity, houses with floors that aren't made of dirt. And in the cities of Nepal, conditions of life for most people are deeply scarred by high unemployment, crowded and substandard housing, and grinding poverty.

In Kathmandu, I see how hard the common people must work to survive. When I go out early in the morning, laborers working at building sites are already hauling heavy baskets of bricks – their backs slowly being twisted by such primitive methods of construction. Vendors, who lug their goods to the main squares and sidewalks, sit in the hot sun from early in the morning to late in the day. Childhood is robbed on the streets where youngsters battle among themselves to find customers who will buy a trinket or offer their shoes for a shine. Women begging in the streets often do double duty, tending to several children while trying to ensure they will all eat that day.

One day we visit someone in a typical Kathmandu neighborhood. It's the middle of the day and people are going about their daily lives, small shops everywhere are selling various foods – vegetables, chickens, and other meats – which will never experience the science of refrigeration. I notice two sidewalk tailors sitting on the ground with what look to be turn-of-the-century (1900 not 2000!) sewing machines, set up to do mending for people. The main streets and side streets are full of dark cubicles, about 10 feet by 10 feet, where people are trying to make a living selling various services or merchandise. Across the way there's a big pile of garbage which a woman is picking through, looking for anything that might be of value. It strikes me again how for the majority of people in the city, the daily struggle to survive is intense and unrelenting.

Nepal has no auto industry and most of the trucks and cars on the streets of Kathmandu are made in India. There are very few privately owned cars, but many forms of transportation to choose from – taxis, bicycle-rickshaws, small motor-cabs, buses, and bicycles.

Everywhere you go there are rickshaws. I am amazed at the men who drive these bikes, how they maneuver through the crowded streets between other bikes whizzing by, past cars and trucks that come within inches. Somehow, they manage to seriously threaten, but avoid running down, pedestrians. Most of these men look to weigh less than 150 pounds – small and skinny by American standards. But they have amazing strength. They can pull their own weight, plus another 500 pounds (three passengers, plus luggage) through streets of unpredictable ruts and bumps. Sometimes, on an incline, the rickshaw will come to a complete stop and then the driver will stand up on the pedals and use all his weight, rocking back and forth to get some momentum to start moving again. Sometimes he will have to get off the bike and put his shoulder into pushing and pulling the load up the road.

From the first day I arrive in Kathmandu, I immediately dislike the inequality of the social relations between the Nepali people and tourists. The poor migrate into tourist areas every day, to try to hustle enough to put food on the evening's table. People are forced into this subservient role where they are constantly asking anyone who looks like a foreigner if they can serve them in any way. Children work the tourist areas, approaching people with the few words of English they've learned: 'five rupees, please.'

The Thamel, tourist part of Kathmandu is different and more distorted than any other part of the city because everything is geared toward the tourist industry. For instance, there are only a few restaurants here that serve Nepali food. The restaurant signs here are meant to appeal to those who travel to 'exotic places' but want the comforts of home. They're designed to entice hungry foreigners who have lots of rupees in their pockets – people looking for Italian, Mexican, Indian, Japanese, or American cuisine. The American bacon, eggs, and toast breakfast is standard fare here and I even come across a café that plays jazz and blues and serves Starbucks-type coffee. The price of a dinner here, by American standards, is very cheap – you can get full on a very good meal for 200 rupees, that's about 3 dollars.

But such prices are totally out of range for local people. A laborer who works on the roads may get 60 rupees for a whole day of backbreaking work, with which he must feed his whole family.

There is a weird mix here of overwhelming poverty and lack of infrastructure, dotted with spots of high-tech development that have been brought in to cater to tourists and foreign business. Most of the Nepalese people in Kathmandu are forced to live hand-to-mouth in impoverished conditions. Meanwhile a rooftop sign on a luxury hotel, perched next to a big satellite dish, advertises 'sauna, massage, restaurant and bar.' Most Nepali homes don't have telephones. But a homesick American in Thamel can send a message halfway around the world in a matter of seconds. When I walk into a small office that advertises 'e-mail,' there are five up-to-date computers and a couple of Americans busy sending e-mail messages for six rupees a minute. Travel and trekking agencies have computers and hotels to keep their clients happy with air conditioning and cable TV. But high schools and colleges in Kathmandu don't have access to anything like this kind of technology.

Unemployment in Nepal is very, very high, and many Nepalese people, especially young men, are forced to go abroad to find work. Most of them go to India or countries in the Persian Gulf. There are 26 million people in Nepal; and it is estimated that as many as seven million live and work in India and around 100,000 work in the Gulf countries, including Qatar, Saudi Arabia, Bahrain, and United Arab Emirates. When I was there, the Nepalese government had just signed an agreement with Qatar to supply 15,000 Nepali workers a year to Doha.

While in Kathmandu I interview several leaders of mass revolutionary organizations – people doing work among women, intellectuals and professionals, students, and workers. This gives me a picture of the Maoist movement in the city and how city people are being mobilized and organized to resist the government and support the People's War in the countryside.

These mass organizations see their work in terms of short- and long-term goals. They lead the fight for better living and working conditions. But at the same time, they wage these struggles in the context of their support for the overthrow of the government. The leaders of these organizations were very clear in their opinion – and many of their organizations make this point in their constitution – that the people's problems cannot be solved until the people get rid of the present feudalistic, corrupt system and build a new democratic system. These mass organizations educate their members in this way of thinking and openly declare that they 'morally support' the People's War in the countryside.

These organizations are all open and legal at this point, but they are all facing increasing government repression. Many people are being arrested, jailed, and killed. But they say this has only made people stronger in their resolve to fight the government. There is a real struggle for these organizations to remain legal and not be driven underground.[1] They have already suffered the loss of many people, but they tell me that, increasingly, many people consider the People's War to be the only hope for real change.

Walking round Kathmandu, I notice high up on many lampposts, red flags with a pen and star insignia. These had been put up by Maoist student organizations in the city.

The student movement in Nepal is very big. All the different political parties have student organizations and when I visit some of the college campuses around the city, I see posters put up by all the different student groups, including the Maoist student organization, the All Nepal National Independent Students' Union-Revolutionary (ANNISU-R). I interview the president of ANNISU-R and he explains the aims of his group and why they support the People's War:

> 'The educational system is a feudal system of education and is not productive, not useful to the people. Many people are not getting an education – which is expensive, privatized, and inaccessible to common children. There is unequal education but equal competition. Good education is accessible only to the elite and ruling classes and the educational system reflects a feudal and imperialist culture ... We can't solve the people's problems through reforms and need drastic, total change, which can only be solved in a new democratic system. For that, we support the People's War. In the long term we see establishing a new people's educational system where there is opportunity for everyone. Education should be compulsory, scientific, and accessible. People should be able to get jobs. Education should have a good link with the production system; there should be no foreign intervention in the educational system and it should be oriented in such a way that students come to know the world and the people of the world. Education should not be a commodity, it is a right of the people. Education should be free and compulsory. It should be oriented to different nationalities and should promote feelings of love for the nation. Policies should be made with and for the needs of the people. There should be freedom

in different academic careers and freedom of speech. Students should have physical labor as well ... Feudal traditions prevail. Most parents don't send daughters to school – there is over 40 percent illiteracy in the countryside. A revolutionary program for education will have a regional balance in education and facilities for education for indigenous children and women.'

The Maoist student groups are very active in the cities and are responsible for 'morally' as well as concretely supporting the fight going on in the countryside. Many students, if they are arrested and forced to go underground, end up leaving the city and joining the People's Army.

* * *

While in Kathmandu I interview Kumar Dahal, president of the All Nepal Federation of Trade Unions. He starts off by explaining that there are two types of workers in the city: service-oriented (hotel, trekking, transport) and production workers. He explains,

> 'Proletarians are a very small percentage in the country as a whole, and the industry that exists is suffering because of globalization. Production workers produce sugar, cigarettes, matches, textiles, etc. Raw materials are imported, for example there is production of rubber goods with imported rubber. There is a carpet industry but this is going down. Foreign intervention in different ways is ruining Nepali industry. Paints, Coca-Cola, Fanta – these factories are totally run by multi-national companies.'

Kumar tells me there are two types of unions in the trade union movement. He says, 'Besides the revolutionary trade unions there are those mainly related to parliament, pro-government and used for votes. They aren't looking to fight for the needs of workers. The main objective of the revolutionary trade union is to raise the professional rights and benefits of laborers and on the other hand join with others to bring socialism to the country to solve our problems. We struggle to benefit laborers and link to the struggle for socialism.'

I also meet with lawyers, human rights activists, and doctors who are sympathetic to the Maoists – and one group of government workers who quietly organize support for the People's War where they work.

Among intellectuals in Nepal things are very politicized, and every political party has its own organization of intellectuals. I have several meetings with groups of intellectuals who organize support for the People's War among their peers. One man tells me:

> 'It is difficult to work among intellectuals in favor of revolution. Many intellectuals work with NGO/INGOs [non-government organizations], whose work is many times related to imperialism, like the IMF. NGOs work to spread Christianity. About 25 percent work directly with some kind of imperialist organization. Many come from the middle classes and also some come from the lower classes in the countryside. Intellectuals who favor the People's War work mainly in the ideological field. They hold seminars and other meetings, classes, etc. The mass line depends on the issue – we talk about ideology, which side to support, discuss philosophy, expose revisionism. We also talk about issues like natural resources, nationalism, elections, treaties with India, opposition to the Anti-Terrorist Act, etc. Around certain issues, we expose what it means, why the government wants to support it. There are processions of intellectuals and we work with intellectuals who have different lines – unite programmatically around particular issues. Most fear they will be killed if they get involved because so many teachers, lawyers, etc. have been killed, lecturers imprisoned, etc. In three years the government has killed 25 intellectuals. Before the Initiation work was not very sharp – everyone can talk about revolution, but no one was doing it. After the Initiation it's easier because lines are sharper, but more difficult because of repression and fear.
>
> 'Many activists are being arrested and detained under a "Preventive Detention Act" which allows the police to arrest people and hold them in custody on the basis of a police claim that the person was going to carry out some illegal act. Most of the time the police use this to just harass and jail people and eventually people are released with no charges. But there have also been cases where people are arrested and then "disappeared."'

A number of the revolutionary intellectuals I talk with characterize intellectuals in Nepal as being a very wavering force. They say that such forces tend to go the way the wind is blowing. If the People's War seems to be gaining and advancing they are enthusiastic and supportive, but when there is heavy repression or setbacks for the revolutionaries, these forces tend to retreat and become scared to do or say anything in its support. But, at the same time, they tell me, most intellectuals in Nepal do not have any faith in the present government and cannot offer any solution other than the People's War and a new democratic society.

* * *

In reading about the Initiation, I had heard about various armed actions taken in the rural areas. But I had not heard much about what happened in the cities.

During my trip I have the rare opportunity to do an extensive interview with Prachanda, the top leader of the CPN (Maoist), who is underground. One of the things I ask him about is what it was like for the Maoist forces in the cities when they started the armed struggle in the countryside. He tells me:

> 'We realized before the Initiation that after the process of initiation there would be a big process of transformation inside the Party. We thought that, possibly, more than 50 percent of our Party members could fall away but that other new comrades and new people would come and join the Party. We thought this might happen. We considered this question beforehand and we prepared mentally for this to happen because there were so many petty bourgeois tendencies, so much intellectualism. We mentally prepared for dealing with the question of how to sustain the People's War after this big leap ...
>
> 'We had a military plan to attack the police force, the landlords, the local goons in the rural areas. But we did not have a big plan for sabotage in the capital city because, at that time, we did not want to create a situation where with one stroke the intellectuals would go away from us. We wanted to sustain their support. We did not want to make the intellectuals in the capital city or other cities run away and stop working with the Party.
>
> 'What happened was that, after one month, we saw a big change in the rural areas. Big changes started in the rural areas. Some people fled. Some new people came forward. Thousands and thousands of people went underground. In Rolpa, in one month, thousands of people went underground. Not only Party members but also masses went underground – in Rukum, in Jajarkot, in Sallyan, in Kalikot. That kind of situation developed. So the process of transformation was very big in those rural areas.
>
> 'But in the cities where there are more intellectuals, the process of transformation was very, very low, and in some cases, we can say,

unsatisfactory. We were not satisfied with the petty bourgeois reactions. One example is what happened just after the Initiation of the People's War, in the capital city, in Kathmandu. There was government repression everywhere. Artists, journalists, professors, lecturers, everywhere, those who sympathized with us were arrested. And at that time, what happened in the city? Wherever we went people said, "You should not stay here, the police will come." There was so much terror among the different sections of intellectuals. For a long time they had been with us. But at that point, they were so afraid, they had so much terror, that even for us, they did not dare to fight, to give us shelter. So for 22 days we had to be continuously on the move in the city.

'But when we got a report from Rolpa, Rukum, Gorkha, Sindhuli, Kavre, the rural areas, there was confidence among the masses and the revolutionary cadres. The sentiment there was, "Yes, we have done a big job. Now new life has started." There was new mass support and mass upsurge in the rural areas. But in the city, the intellectuals were vacillating so much, they were so terrorized, and we saw that this was a question of class. Which class thinks that now we are taking destiny into our hands? This was the situation, and so we had to wage ideological struggle in the cities.'

I know that the Maoists do work among workers in the city and I ask Prachanda what role the proletarian forces played in the cities at the time of the Initiation. He says:

'During this tough time there were the forces in the city – laborers, workers – who helped the Party, saved the Party. Workers from our All Nepal Trade Labor Organization helped us a lot. They were not so terrorized. Their feeling was, "OK, this is a new thing." And another important section was women – this is very, very important. Women in Kathmandu were the other force who, in that time of terror, boldly supported us and gave us shelter and helped us move around. The women helped us at that time. There was also help from students because we had good organization among the students all over the country. And at that time the students were also not so afraid. They felt enlightened, that this was a new thing for Nepal that our Party had done. This kind of thinking was there among the students. So it was mainly laborers, women, and students who helped us. But the intellectuals, who had a lot of knowledge of philosophy, theories, etc., these people were vacillating so much they could not help very much.

'And after one year, we saw further big transformation in the rural areas. Thousands and thousands of mass organizations were built up, and in new areas the Party's influence spread and new organization developed. Some petty bourgeois revolutionaries fled in terror to India, to Arab countries and other places. Others stayed strong. And at the same time, in the rural areas, there was a mass upsurge of women, and thousands of full-time cadres came forward. People who we did not know beforehand became heroes, really. Just one year after the Initiation, for one month, I was in Rolpa, Rukum, Jarjarkot, Sallyan, and I saw, and our Party saw, that a new thing had developed. The people were not only fighting with the police or reactionary, feudal agents, but they were also breaking the feudal chains of exploitation and oppression and a whole cultural revolution was going on among the people. Questions of marriage, questions of love, questions of family, questions of relations between people. All of these things were being turned upside down and changed in the rural areas.

'We came to understand Mao's vision that the backward rural areas will be the base of revolution – the real base of the revolution. We saw in Rolpa, Rukum, Gorkha, Sindhuli, Kavre the seeds of the new society, the examples to inspire people. Everywhere in the country, in the revolution, the masses feel proud of their Rolpa and Rukum. And we see, at the ground level, on the mass level, that the transformation process is not only in the Party and mass organizations, but among the masses as a whole. The chains of feudal oppression, mainly feudal relations, are breaking.'

6
General Strike in Kathmandu

On Friday I get up early, anxious to witness my first *bandh* (general strike) in Nepal. The Newa Khala (Newari Family) an association of the Newar nationality, an oppressed minority in Nepal, has called for a Kathmandu Valley-wide *bandh*. The newspapers report that this is a protest against increasing 'terrorist activities, unemployment, and price hikes' and 'discrimination based on race.' (By 'terrorist activities' they mean *government* repression against the people.) The news also reports that the government Home Ministry has claimed that since Maoists are supporting the *bandh* it may not be 'peaceful,' and so police have already arrested more than 80 people accused of organizing it.

Today the streets look very, very different because most of the shopkeepers are observing the *bandh* and hardly anything is open. I hook up with someone who has been showing me round and we go out into the city on foot. We walk over to New Road, a major thoroughfare with lots of stores. I'm told that some people are exempt from the *bandh*; these include rickshaw drivers, small sidewalk vendors, and pharmacy owners. We go through an area which yesterday was jam-packed with pedestrians and heavy traffic. Now there are still lots of people about, but hardly any traffic, only bicycles, motorcycles, and rickshaws. For Kathmandu, this looks very strange. For once, it's almost safe to cross the street.

At a couple of major intersections I see, for the first time, the Nepali riot police – the commandos, as they're called. A small group are standing in the middle of the street wearing padded armor and helmets, carrying batons. Later, I see some commandos who are even more heavily fortified. Some have rifles and they are all wearing padded armor, not only on their torsos, but also on the front of their legs so they look like weird armed hockey goalies. My friend tells me that there have been hundreds of arrests, most in the last 48 hours, in connection with the *bandh*.

I take a two-hour walk in the afternoon to Lalitpur (Patan) just over the Bagmati River. Along the way, almost all the shops are closed and there are still no cars on the streets. If I had just arrived in Kathmandu for the first time, I might have thought things were quiet but 'normal' – it would not be readily apparent that a mass political protest is taking place. I have heard that in the cities, anti-government sentiment is widespread and that there is a lot of support for the People's War. But there is also heavy government repression against anyone who openly sympathizes with the Maoists. In such a situation the *bandh* is a form of mass protest so it's hard for the government to single out and punish individual merchants. That's what makes it easier for people to participate.

On the way back to Kathmandu, I pick up a copy of the *Kathmandu Post*. Today's editorial, about the 'growing problem of the Maoist insurgency,' points out that,

> The insurgency itself has its roots in the economic hardships, poverty, deprivation, and exploitation that has been the lot of the majority of Nepalese. The fact that successive governments have failed to address these problems in the post-democracy era has further helped fuel the fires of insurgency in many areas of the country. Thus, it was only natural that many people should take to armed struggle against the glaring inequities and social injustice when the government failed to deliver or make any perceptible impacts felt in the backward regions of the country. It is therefore not at all surprising that insurgency, which was practically unheard of in the country prior to the restoration of democracy, should attain such threatening dimensions.[1]

This gets me thinking about how 'economic hardships, poverty, deprivation and exploitation' are especially severe among the dozens of minority ethnic groups in Nepal, and how this must be a motivating factor for these groups to support the People's War. The program of the CPN (Maoist) guarantees 'equal treatment to all nations/nationalities and languages' and states that within the framework of the New Democratic system they want to establish 'all forms of exploitation/oppression of the oppressed nations/nationalities shall be ended and they shall exercise their own autonomous rule in the land they inhabit.'[2]

After the *bandh*, I want to find out more about Newa Khala, which organized the *bandh*. I arrange an interview with Dilip Maharjan, the chairperson of Newa Khala. After introductions and some words of greeting, Dilip Maharjan starts by explaining the goals of Newa Khala, established only nine months earlier:

'Our objective is to classify the problems that indigenous Newars are facing in Nepal and to organize different struggles to solve these problems. We especially organize Newars to make them aware of the oppression and suppression by the state power. We mainly concentrate on language, religion, cultural values, the economic downfall of the Newars, and social domination by other religions – mainly the religions of the ruling class. The state power oppresses Newars in many ways. To understand this we need to go back to the history of Nepal. Nepal was divided into tiny kingdoms and was then united by ancestors of the present king with the formation of the central feudalistic government. The Newars were oppressed by the central state power. Before that Newars had their own kingdom, language, and culture. They were sound economically and culturally and had established a trade and business role in the trade route between China and India. And so they had a high status in the valley.

'The Newars have their own dignity and identification but because of the centralized government the Newars have been discriminated against. The Aryan and Hindu religions are considered to be superior and the government treats these people much better. In offices, schools, everywhere, the Nepali language is the national language and Newars have been forced to read in Nepali and can't read in our own tongue. Most Newars belong to the Buddhist religion but this is a Hindu state. Hindu religion, Nepali language, Aryans – all this is seen as superior. In this way, by law and in practice, the Newars are being oppressed and face difficulty in maintaining our identity. The Newars are the indigenous people of the valley but are being displaced by Marwaris (the business caste from India). Newars are not getting jobs, they have no opportunities to join the army and police or participate in politics. In this way Newars are discriminated against by the state power and we want to make this known and fight against it and fight for the security of our rights. The Newars are being oppressed by feudalism, capitalism and the intervention of imperialism.

'In Kathmandu city there was a decision that people could use the Newar language to solve problems – that if someone doesn't know Nepali they can appeal in their own language. But about one year ago, this was overturned by the Supreme Court which said we cannot use our own language and must use only Nepali. Now we can only do appeals in English or Nepali. Also, Newar children don't get any opportunity in school to learn their own language.'

Dilip then explains why they called for the *bandh*:

'Our vision is that Newars won't have the right of self-government and self-determination and be able to preserve our language, culture, religion and traditions with the present state power. We called the Valley *bandh* program around 15 demands to the government with our main objective to pressure the government to meet the demands of self-government and self-determination. If the central government discriminates against the Newars, then the Newars should have the right to their own government. One of the demands is for a secular state with all religions having equal rights. Our press named different parties and individuals in support of the *bandh*, including CPN (Maoist), United Front Nepal, NCP (ML), and Mashal and Unity Center. Different forces united to support the strike. More than 80 percent of the people in the valley are Newar.

'For the strike, we prepared many pamphlets and 100,000 posters were distributed. We went house-to-house and met with community leaders to build support, held programs to build support, our leaders made appeals in interviews, etc. Other groups issued letters in support, including indigenous groups. More than 500 people were arrested – there could be even more, it is hard to assess. Our members and activists were arrested passing out leaflets and posters. There was government terror from the beginning, and leaders had to go underground. There have been widespread, indiscriminate arrests in the community to create an atmosphere of terror. People are still in jail.'

I ask Dilip if there was support for the strike among other ethnic groups and he answers:

'There are fraternal relations with other organizations of oppressed nationalities. Our view is that only by being united can we get success. And we want to build mutual help in each other's struggles. We make clear that our vision is not according to caste or religion or community, but it is by class. Our vision is to look at this problem in a class way. We look at the oppressor caste people in a class way. We must protest against a kind of oppression that comes in the name of Indo-chauvinism. But this chauvinism exists among indigenous people and we have to struggle against this. Some Newars belong to oppressor classes. We are with the oppressed class. Chauvinism is expressed through the Hindu caste. We must have good relations with the oppressed and must make bad relations with the oppressors, even within the oppressed ethnic groups.

'There is big support for our demands and the *bandh* has opened the path for our organization to go to indigenous groups and say these are the demands of the people and we must fight. This was the first time a general strike around demands of the Newars has been organized in the valley. It was an historical event.'

Carrying the Story Forward: The Problem of Disinformation

In the years since my trip, I have continued to closely follow developments and monitor coverage of the situation in Nepal, including news reports, articles, and analysis from publications and websites clearly unsympathetic to the Maoists. It has always been hard to obtain reliable information about what is going on in Nepal. But this was especially true after the government declared a State of Emergency in November 2001 through 2002. While the Royal Nepal Army waged a 'search and destroy' campaign in the countryside, a disinformation campaign was launched against the People's War.

Newspapers sympathetic to the Maoists were raided and closed down, and their staffs arrested. Even the mainstream press in Nepal was targeted; editors and writers were picked up and interrogated for simply quoting Maoist leaders in their publications. Journalists were arrested for protesting the arrests of fellow journalists.[1] In the first nine months of the State of Emergency, 130 journalists were taken into custody.[2]

Meanwhile the Ministry of Information and Communication issued strict guidelines for the media with regards to news about the People's War. The articles the directive stipulated could *not* be published were: 'anything that aims to create hatred and disrespect against His Majesty the King and the Royal Family'; 'anything likely to create hatred against Royal Nepal Army, police and civil servants and lower their morale and dignity'; 'any news that supports Maoists, including individuals or groups'; and 'any matters that aim at overthrowing the government.' What could be published fell into three categories: 1) 'news that exposes criminal activities of Maoist terrorists'; 2) 'news regarding bravery and achievements of Royal Nepal Army, police and civil servants'; and 3) 'official news that comes from His Majesty's Government and the official media.'

As part of the media crackdown, journalists were barred from going into the battle zones, and the RNA said the media should seek permission from the army's Information Department before publishing any news about military affairs.[3]

The State of Emergency formally ended in August 2002. But there continues to be widespread disinformation about the People's

War aimed at vilifying and discrediting the Maoist insurgency. Unfortunately, human rights groups have also contributed to the proliferation of disinformation, sometimes uncritically repeating reports issued by the Nepalese government or pro-government publications in Nepal. One article in the *Kathmandu Post* reported that Maoist guerrillas were going into battle drunk and on drugs, with pockets full of condoms.[4] I never saw independent reports of this in any other publications and it seemed like a ridiculous claim from everything I have read about the guerrillas. But there it was in the biggest newspaper in Nepal and out on the Internet for millions of people around the world to read.

* * *

The following news item appeared in the 'World Briefing' section of the *New York Times* on January 16, 2003:

> **Nepal: Conflict Toll on Children.** Two thousand Nepalese children have been orphaned, 4,000 displaced and 168 killed in the Maoist insurrection that began in 1996, a Kathmandu-based human rights group reported. The group, the Child Workers in Nepal Concerned Center, said more than 10,000 children had been denied access to education because of the insurgency. The center's president, Gauri Pradhan, expressed concern about reports that children have been recruited into the Maoists' army.

In three short sentences the *New York Times* takes statistics about the total number of deaths, from both sides, and then implies that all these deaths are caused by one side – the Maoists. Four statistics are given in the article: 2,000 orphaned, 4,000 displaced, 168 killed 'in the Maoist insurrection' and 10,000 denied access to education 'because of the insurgency.'

In fact, the overwhelming majority of the children who have been killed in the insurgency in Nepal have died at the hands of government forces. And most of the children who have lost their parents have become orphans because their mothers and fathers were killed or rounded up and jailed by the government.

As for Maoists being responsible for 'denial of education,' here too the facts need to be examined. Maoist students have organized strikes, shutting down schools throughout the country – but they are doing this not in opposition to education but to press their demands for equal opportunity to a free and relevant education. There are instances in which schools have closed down in areas under guerrilla control because reactionary teachers have left these areas. But the Maoists also build and run new schools in their base areas – schools that are not likely to be counted in official government statistics.

The larger point is this. A civil war is raging in which two sides are fighting it out to determine the future of Nepal. The whole country is affected, and there is much turmoil in the countryside as a result of the insurgency and the government's attempts to crush it. Great numbers of people have left their villages in the wake of vicious counterinsurgency campaigns mounted by the government. At the same time, there are people leaving areas under Maoist control, and this happens for a variety of reasons: some of these people are reactionaries tied to the government and know they are hated by the local populace; others leave because they do not support the Maoists; and some are simply seeking to avoid the hardships and conflicts of the war.

The *New York Times* claims that '4,000 people have been displaced' by the insurgency. But a more compelling fact – recognized even by those fighting against the Maoists – is the support and active involvement of *millions* of poor peasants in this revolution.

* * *

Disinformation has long been a basic tool of US foreign policy. In 1964, the US fabricated the 'Gulf of Tonkin incident,' claiming the North Vietnamese had fired on US naval ships. This was the pretext and justification for giving President Johnson executive power to expand the US war on Vietnam. In the 1980s, US Army 'psy-op' units broadcast radio and TV programs into Nicaragua intended to undermine the Sandinista government. During the 1991 Gulf War, the US released numerous false news reports, like the completely fabricated story that Saddam Hussein's forces had taken babies out of incubators in Kuwait, leaving them to die. And of course, the 2003 war against Iraq was justified by the bogus claim that the US knew Iraq had 'weapons of mass destruction.'

There is also a long history of the US seeking to discredit revolutionary forces to spread confusion about the actual nature and aims of revolutionary struggles. This is done to justify brutal counterrevolutionary measures, to turn allies and potential supporters against these struggles, and to undercut popular outrage and protests against the human rights abuses carried out against revolutionaries and their supporters by regimes supported and propped up by the US.

The mainstream press in the US has mainly taken the approach of ignoring and censoring news about the People's War in Nepal. But increasingly, and especially after September 11, there have been more systematic efforts to spread lies about the Maoist insurgency. Here are some examples.

On January 20, 2002, *USA Today* wrote about Nepal: 'The rebels have attacked civil servants, including schoolteachers, whom they have maimed by cutting off their hands.' There have also been reports in the mainstream press about 'execution-style killings and maiming of scores of teachers.' One article I read claimed a teacher was killed simply because he went against the Maoist demand that teachers stop teaching the archaic Sanskrit language. And Amnesty International issued a statement saying that Maoists killed Mukesh Adhikari, a teacher who was also a member of Amnesty International, because he was a member of a teachers' association affiliated with the ruling Nepali Congress Party.

Such claims make it seem as though the Maoists in Nepal are ruthlessly attacking teachers that disagree with them and are somehow against education. Such stories would have people believe that the Maoists are fanatics like the Taliban who cut off people's hands and are against progress. But this contradicts everything I know about the actual character of the revolutionary struggle and who is targeted in the People's War.

In a letter to Amnesty International, the International Department of the CPN (Maoist) expresses deep concern at Amnesty International's 'one-sided and exaggerated reporting.' The letter points out that 'the royal army has been massacring dozens of unarmed and innocent persons every day in fake encounters and the media and human rights organizations have been strictly barred from visiting the sites for independent verifications.' The letter then says:

> As regards to specific cases of execution of some persons who also happen to be teachers by Maoists, you seem to give the impression that they were executed because of their political affiliation to Nepali Congress Party or its mass organizations. In this regard, we would like to once again reiterate that CPN (Maoist) is against violence per se. However, having said that, we would also like to remind you that we do not hesitate to use violence as a defensive measure when the state uses its coercive instruments against the people ...

> When any person, whether a teacher, or a student or an intellectual, or manual worker, or anyone irrespective of his occupation or political affiliation is found repeatedly abetting to this fascist state in carrying out massacres of the people, he is punished according to the revolutionary law.
>
> Mukesh Adhikari, who also happened to be your local representative, was one such person with long record of anti-people activities and proven charges of an informer leading to the massacre of a number of innocent people by the royal army in Lamjung.

When I was in Nepal I heard stories of individuals who had been targeted for execution by the guerrillas. Sometimes these individuals were members of the ruling Nepali Congress Party or of another party in the government. But what's important to understand is that these people were not targeted simply because they were members of these parties, or just on account of their holding government office. They were targeted because of specific acts they were responsible for or had taken part in, and usually repeatedly, that had resulted in the death or torture or arrest of Maoist guerrillas, party members, and sympathizers.

Those who fear the success of the People's War in Nepal will spread the most outrageous stories, sometimes even knowing that these will be quickly disproved. They hope that if they put out enough lies they can discredit the Maoists and generate broad public opinion against the revolution. And by using the 'terrorist' label, they want to declare 'illegitimate' and out of bounds certain struggles – even a profound social upheaval like the People's War in Nepal, which clearly enjoys the support of millions and millions of poor peasants.

7
Land in the Middle

The Middle Region of Nepal lies between the Kathmandu Valley, where the government rules from the capital, and the Western Region, the heart of the People's War. The biggest city here is Pokhara, where lots of tourists go to view the famous Anna Purna range and start out on treks. This region also contains the large Terai area. This narrow strip of plains land along the southern border with India includes another major tourist area, the famous Chitwan National Park.

People tell me that in the Middle Region there are more landlords who own a lot of land. In the hill areas, most of the peasants still own small plots of land, but there are also some who are landless and work for landlords, getting paid monthly or yearly in money or crops.

In the Terai, the peasants mainly work for daily wages. Here, the soil is highly fertile and there are abundant water sources, which permit the cultivation of a wide variety of crops. But about 20 percent of the people are landless and others have only very small amounts of land.[1]

The Terai was heavily forested until the 1960s, when the government and landlords began to clear the area to deal with the problem of malaria,[2] so the land here does not have a long history of being owned and cultivated by the peasants.

In the Terai, the Maoists have seized land from landlords and also freed some *Kamaiyas* – bonded laborers who had been 'sold' to a landlord or forced to work unpaid in order to honor a debt.

I heard the story of one man who had been working as a bonded laborer for 50 years. He had been stolen from his family when he was only nine years old and was sold to the landlord for 20 kilograms of millet. The Maoists here say they led nearly 1,000 people to go to the landlord and hold a mass meeting. The mass revolutionary organizations exposed the kind of exploitation done by this landlord and called on the people to take mass action. The people demanded that the landlord either pay the man for 50 years of labor or give him property as if he were his son. The landlord accepted the second proposal and was forced to give up 0.8 hectares of land and about 400,000 rupees.

All round the world, in poor Third World countries, landownership is a crucial question for hundreds of millions of peasants. And this is a central question in the People's War in Nepal. When I interview Prachanda, I ask him to talk about how the CPN (Maoist) looks at the process of revolutionizing production relations in the countryside. He tells me:

> 'Basically, the character of this revolution is agrarian. But the situation in Nepal is not classical, not traditional. In the Terai region we find landlords with some land, and we have to seize the land and distribute it among the poor peasants. But in the whole mountainous regions, that is not the case. There are smallholdings, and there are not big landlords. Therefore our main plan in those areas is to develop collective farming and revolutionize the production relations. How to develop production, how to improve productivity is the main problem here. The small plots of land mean the peasants have low productivity. With collective farming it will be more scientific and things can be done to improve productivity. But we cannot do this collective farming instantly. In terms of landownership, it will be private ownership by the peasant. But the production process will be collective. We are trying to do this in our regions. And, mainly in our developed regions, collective farming has already been established.
>
> 'In the developed areas we have already made a plan and started, in some areas of Rolpa and some areas of Rukum and some areas of Jarjarkot and some in Sallyan – less in Sallyan, mainly in Rolpa, Rukum, and Jarjarkot. First we seized some land from landlords who live in Kathmandu, and from usurers and such types. We seized that land, but we did not distribute it to the peasants, because to distribute that land piece-by-piece to peasants will not work, will not help to develop their livelihood, their economic level. So number one are the kinds of land seized from landlords, usurers, etc. Number two are other lands, like public land which can be cultivated. And number three are lands owned by peasants. These are the three types of lands that are there. When we seize the land from landlords, that land will become collectively owned, there will be collective mass ownership. That land will be the land of the masses, and all the peasants will work on that land, and the earnings from that land will be the property of that locality.
>
> 'The return from that land will be the collective funds of the masses, used for the needs of the masses of that locality. Up to this point we have done it like this. And the fallow land or public land which can be cultivated – we are trying to cultivate this land collectively and distribute

the return to the masses collectively. Collective distribution means according to what percent of the work has been done, according to the number of hours worked. How many hours' work a particular family did on that land, the return will be in that percentage.

'Where our mass base is strong and the masses are in the process of struggle, we are starting to have collective farming. Private ownership, but farming collectively. This has already shown effectiveness in the process of production.

'Animals, tools, land – according to the land, according to the tools they use and work hours, in that percentage, according to the percentage of the work done, they divide the production. In this way we can raise the quantity of production. This is what we are doing in the developed areas. But in less developed areas, in the Eastern Region and the Middle Region, we are trying a kind of system that is not exactly an exchange of labor power. But like during the rainy season, if you have less manpower or your labor power is not sufficient and you cannot do well in cultivation, then other peasant families are there to help. My family will help you, and your family will help me, and we will help him. This kind of tradition is there in peasant families. These kinds of traditions were there, and now we are developing this tradition in an organized way. And in a more organized way we are starting to develop different kinds of collective farming and measures that lead to collective farming. We are trying to organize this system of farming, and it enables the peasants to achieve a kind of unity among them. They are doing all these things to break the chains of feudalism, and it is a school of cultural transformation. When all our families work together, eat together, sing together, dance together, then it is more communal.

'In the Terai, up to this time, we haven't had a strong mass base, there is not a strong struggle. There is guerrilla action going on in the Terai, in the plainlands. There are big landlords, there is king's land, queen's land – so many big bourgeois lands are there. Up to now what we have done is seize the grain of landlords. We are not yet able, in the Terai, to seize the land. But we are able to seize stored grain. This enables the masses to understand the importance of the People's War, the importance of the revolution. Gradually, they are coming to see, "Yes, this is ours." And so we are also developing a mass base in the Terai region.'

* * *

Compared to areas I visited in the east and west, the Middle Region seems a lot less remote – closer to cities and major highways, more invaded by tourism, and more vulnerable to the movements and mobilizations of the police. There is also a lot of interaction with and influence from India, which is right across the southern border. Some of the guerrilla zones here are closer to small urban areas and more linked up with revolutionary work being done in the cities.

I visit a small city in the Terai where people arrange for me to meet with representatives from four People's Army squads, mainly commanders. These guerrillas have taken the risk of coming into the city from the countryside and we have to travel cautiously to the meeting place.

The squads these guerrillas lead operate near an area in the west that is becoming a base area and so, they tell me, their squads have an important duty to support these western areas. In addition to carrying out actions in the Middle Region, they also do support work, providing couriers and arms and sending 'full-timers' to the west.

Unlike the rest of Nepal, the climate in the Terai is very hot and humid, with lots of mosquitoes. We are packed into a small storeroom, sitting on the floor, knee-to-knee. There is one fan which provides a small breeze of relief. But the guerrillas only turn it on for brief spurts because they don't want the noise to attract attention. One squad commander starts by recounting some of the first actions taken by his squad:

> 'Several squads have been working since the Initiation and one other was established in the countryside a year ago. We concentrated initially on propaganda work, wall paintings, postering, and processions. We had other goals like seizing arms. But in the beginning, we weren't able to do this. After one year, we were able to do a seizure of arms. And now, during the last three years, our squads have done more armed seizures than other actions.
>
> 'There was one man who didn't pay wages to his laborers and took high interest on loans. When the Initiation happened he bought a 12-bore rifle from India and threatened people with it. He also sexually harassed many women and so we decided to take action against him. There were twelve in our squad and at 9:00 pm we went to where he was, in the forest. We divided into four groups – an assault group, seize group, reserve group and defense group. We did a surprise attack. He was sleeping with his gun and more than 60 workers of his were in the area. When we attacked he got a broken hand, and we were able to seize his weapon. The workers tried to attack the squad, but we told them, we are Maoist guerrillas, and we explained all the ways their boss

was exploiting the people. We said to the workers, "We are fighting for you, not the goons," and all of them fell silent.'

The squad commander then tells about actions they have taken to get land:

'Another action we did was a land seizure. We seized more than three hectares where the government had planted trees. We took out the trees and the boundary and distributed the land to the landless, who now live on it. We also seized crops from some landlords and distributed this among the people. Then the party decided to target this Indian comprador and landowner. He is a citizen of Nepal and India and had a number of servants working on his land. We convinced a lot of the poor peasants who worked under this man to attack him.

'So at night, at 10:00 pm, seven squad members encircled his house, together with nearly 100 peasants. He was sleeping on the second floor and we called out, "We are the forest rangers." He woke up and came outside on the railing but didn't come downstairs. Then we gave a *real* introduction, saying, "We are Maoist guerrillas with the peasants." We gave a speech exposing his bad role and four squad members went to seize his harvested crops. We broke down the wall of his storehouse and all the peasants went inside, shouting slogans.

'We distributed the crops and people took as much as they could carry. But even this was only part of all his wealth. The slogans we did this under were: "Long Live the Peasants' Movement," "Long Live New Democratic Revolution," "People's War is Continuous," "Long Live Peasant Organizations," and "Smash Landlord Ownership." We came back to the house later and took some high quality rice. Then the police came round to people's houses to look for this rice, but they couldn't find anyone with it. At first they didn't arrest anyone, they just terrorized people. But then, within 24 hours, six people from among the masses were arrested.

'We carried out another action with a twelve-member squad in which the house of a landlord was blasted with a bomb. When the squad was returning from this action, about 150 people surrounded them. The people thought we were robbers. So we gave a speech telling the masses what we were really doing. But there were seven people in the crowd who were working for the landlord as a security force, and they called on the people to "unite against the outsiders" to defend "our place." They said, "The Maoists are from the hillside. And if they come here they will displace us." This was in the Terai area, the plainlands, on the border of

India, and the goons raised the differences between those in the Terai, who are originally from India, and other nationalities. The squad tried to control the situation by telling the truth – that everyone should support the People's War. But we could not even be heard because everyone was shouting. The squad was not successful and decided to leave.'

As the squad retreated, the crowd angrily followed them and the guerrillas had to set off a grenade to make their escape.

'The Party had not anticipated that the people would respond like this and so only a small squad had been sent to carry out this action. And this was the squad's first experience. Afterwards, we tried to make sense of this situation, so we could rectify things with the masses. And after some discussion, we came up with two causes/contradictions for what had happened. One was the lack of understanding of the political situation among the masses. And secondly, there is the misunderstanding between hillside and Terai peoples. There are cultural and language differences between hillside and Terai peoples. So we went to people's houses where the incident took place and talked with them. And we found out that in fact, the people were not against the Maoist movement. They told us that they reacted the way they did because they were afraid of repercussions from the security forces and the landlord. Still, most of the people in this area are wavering in their support, depending on how strong the People's War is. So the Party went among the people and worked even harder to build revolutionary mass organizations.

'So far, we have threatened three landlords and forced them to return land to the peasants. Landlords have security guards, weapons, etc. and are very strong, and it's more difficult to attack them than to raid police stations. In this area many of the peasants are landless and have to work on the landlords' land. Some have even had to sell family members to the landlord. Those who work on the landlords' land get just subsistence food and clothes to live, others get some crops. So now the people want to work as free laborers and get wages from the landlord.

'In this mid-region, only 2 percent of full-timers in the Party are people from the Terai. But the Party has announced the right to self-determination and the guarantee of autonomous regions for the oppressed nationalities. With this program, the Tharu nationality people in the Terai are being organized into the revolutionary organization, "Tharuwan Mukti Morcha" [Tharu Liberation Front].'

One of the squad members in the room is a Tharu from a landless family, and tells me, 'When I joined the Tharu Liberation Front, I came to know of the division between the landowners and landless and that the problem of landlords and peasants can only be resolved through class struggle. During this time I was working for a landlord and participated in the liberation movement. We demanded land and more wages so the relationship with the landlords was tense. In this way I joined the Party and participated in actions and got knowledge from the Party.'

There is only one woman in this group and she is eager to speak. As soon as she starts to talk, I can feel her passion, even though I can't understand her words. Her gestures and facial expressions radiate toughness. My translator seems almost transfixed by what she is saying and momentarily forgets his job, so I have to nudge him to get him to tell me what she's saying:

> 'I am 19 years old, unmarried, and went to school through eighth grade. I have been a full-timer for two years. My mother and father are in mass revolutionary organizations. I was a dancer and singer who sang revolutionary songs, but at first I didn't really understand them. I learned them when a cultural troop came to our village. After the Initiation, goons forced four people to go underground and it was dangerous in our village, so I was a little afraid. But I started to act as a courier for the Party. When the police tried to capture me, I left the house and went underground. Then I learned the revolutionary politics of the Party.
>
> 'I knew a little about the revolutions in Russia and Peru. And I knew that women there had participated with guns on their shoulders. After some time I became a squad member and picked up the gun like those women.
>
> 'I had my first experience of giving physical punishment to a bad element. He was a usurer who raped his own auntie and I was very happy to see a number of women take part in this attack. I stood guard with a rifle on the bank of the river.
>
> 'This was the first time I was armed, so I felt more responsibility. And at the time, I thought about Jenny Marx [the wife of Karl Marx]. When we caught the usurer, some neighbors came to see the action and the squad commander gave a speech exposing the target and all the masses supported the squad. I was armed and all the village women were surprised to see a woman like this. The commander said to the women, this bad element raped his auntie so the women should hit him. So I kicked the bad element and blood came out of his mouth.

This was the first time I felt how good it is to hit the class enemy. All the women threatened him and told him that if he rapes anyone else, they will not leave him alone.

'Now that I'm with the squad, I feel a responsibility to fight the reactionaries and to liberate women, along with all the masses of people. I feel more responsibility to overthrow the reactionaries and make a new Nepal. I see how many women are attracted to the People's War and I'm going to work to recruit more women into the People's Army.'

8
Hope of the Hopeless in Gorkha

One of the most interesting parts of my trip to the Middle Region is in the Gorkha district. The area we go to is very poor and the houses are very primitive and small. Instead of clay, many of the houses are simply platforms made of wood with dirt floors and straw roofs. Sometimes there aren't even any walls.

We arrive at the first village late at night and after a quick meal I get a chance to talk with two leaders of the CPN (Maoist) in this district, M.B. and B.K. There's only a single candle cutting a narrow and flickering swath of light, and it's a bit of a struggle to write clearly in my notebook. But in the dim light I can see, as well as feel, their enthusiasm.

A lot of the peasants here own small plots of land. But in Gorkha, compared to other areas, there are a lot more landlords and landless peasants and the guerrillas have led a lot of struggle to seize and redistribute land. B.K. says:

> 'One contradiction the peasants face is with small agricultural banks. They have to pay high interest on loans, and many times the bank ends up taking their land away and they become landless. To solve this problem, we told the peasants not to pay the banks for their loans. We attacked one bank and destroyed all the loan documents, so the peasants were freed from their loans. A second contradiction is between peasants and landlords. We dealt with this by implementing a policy of land to the tiller. Land was seized and distributed to the peasants. The third contradiction faced by the peasants is with individual usurers, who give loans with high interest. This problem was solved by destroying documents for these loans.
>
> 'There are other contradictions as well. For example, in the name of religion, God and the royal family, the priests and government own property that they give to peasants to farm on a contract basis – the peasants work the land for wages. The peasants grow a lot on these lands, but don't get very much. So the Party has led the struggle to seize these lands and distribute them to the peasants. There are also some people who file false claims against the peasants and steal land through the courts and the Party has captured such people and brought

> them before the masses for public punishment. These people are made to confess their crimes and return the land to the peasants. We have also been able to seize some land from the landlords and distribute it to landless people. The rice we ate tonight was grown on these lands.'

The two men tell me that when the People's War targeted agricultural banks, landlords, usurers, and local reactionaries, the peasants saw it was possible to take up arms against those who had been oppressing them. People were then emboldened to support the revolution and started coming to the Party to solve disputes over land, fights, etc., instead of going to the government. M.B. continues:

> 'After the Initiation we were able to seize arms from a bank security guard. We also seized arms from the Nepali Congress Party and a Nepal Bank branch. After this action, two Party leaders were arrested. But whenever the people have felt defeated, the Party has given us a new plan and we have been able to make a comeback. Now, many young men and women are lining up to join the People's Army.
>
> 'After these advances by the People's War the reactionaries got very aggressive in repressing the people. But we were able to carry out a big action against this landlord, who was also the district treasurer of the Nepali Congress and a member of the government District Development Committee. We did a surprise attack on him and captured his money and all his belongings. This man had committed many crimes against the people. Like he would charge huge interest for loans and do things like take away people's buffaloes when they couldn't repay their loans. He took 300 buffaloes from the people in this way. So the masses of people were very happy to see him punished.'

B.K. tells me that even though there was a lot of government repression throughout this whole period, they were able to build organizations among the peasants, students, and women. And they also carried out a successful campaign to raise funds for the Party and the People's Army. He says: 'Through this fundraising campaign, we were able to assess the political support and views of small merchants, teachers and low-level government workers. We learned about their views in this way and found that, in fact, the overwhelming majority of them are sympathetic to the People's War.' He also says that as they gained strength, the reactionary forces in the area became more isolated:

'As for the Nepali Congress and UML, they cannot come into the villages now, and they have all left. None of them dares to come into this area to campaign for the elections. In two districts here, there are about 800,000 people, and about 50 percent support the People's War. There is especially more support among the women, and in this area the women make up about 30 percent of the People's Army. Now, so many people want to join the mass organizations and the Party that we don't have enough organizers to meet the demand.

'Today when I meet the martyrs' families and the relatives of those who are in jail, I see that they are not discouraged. They tell me to work to advance the revolution and they are involved as much as they can to give the People's War support. And even though the police are repressing the people, the masses of poor people are not discouraged. Among the better-off farmers there is some discouragement though. And so our plan is to go mainly to organize the lower caste and landless people. These are the people coming in large numbers to join the People's War, this is the main force of the People's War. The main thing is the political line and ideology, and we give political education to people. In the Party, everyone goes through political education classes and there is discussion of political documents on Marxism-Leninism-Maoism, philosophy, politics, history, economics, etc.'

The next day, I spend many hours talking to K.C., another leader in this district, and he gives me a sense of how hard it was to make the transition to armed struggle. The Party had to go through a lot of changes in the way it functioned and also had to deal with many setbacks because of heavy government repression. Many people were killed in this area, and this put a lot of demands on Party members at every level. K.C. begins by telling me about the difficulties they encountered right before and at the start of the People's War:

'In the past, the Party members here were mainly more educated intellectuals who could not leave their jobs and be full-timers. And when the Central Committee decided to start the People's War, this presented us with a problem. Before the Initiation, we held District Committee-level meetings with all the members and united with the decision to go to armed struggle. We decided it was necessary for all the members to become full-timers, but most said they didn't want to do this. So the Party decided to dissolve this District Party Committee and form a new District Committee made up of dedicated persons and young people who had participated in the class struggle and were willing to do revolutionary work full time.

'Then, at the time of the Initiation, some Party leaders in this district dropped out. They said they agreed politically with the plan to start the armed struggle, but didn't think they could deal with the difficulties of being underground. The rest of the members of the district leadership all went underground. After this we faced new difficulties. The new young comrades were dedicated to carrying out the armed struggle but had not established their leadership. So the veteran comrades had to learn to lead the new young leaders.'

K.C. says that as they led more struggle among the people, this gave the leadership and cadres confidence to start the armed struggle. But, still, a big contradiction was that while there was a lot of support among the people, there were only a small number of Party leaders. This was the situation leading up to the start of the People's War. Then on the day of Initiation:

'On February 13, 1996 at 3:45 pm, we succeeded in raiding an agricultural bank, targeting it as a symbol of imperialism. This was a new experience for us as we had no practice in armed struggle before this. The bank is run by the government and involved in carrying out IMF [International Monetary Fund] and World Bank loans and policies. We took all the documents there and burned them.

'The repression was very heavy in the weeks after the Initiation and people had to stay hidden in supporters' houses. Up to the end of February, no one was killed. But about three dozen people were arrested. Then, on February 27, the police went to arrest the headmaster in one village. The students opposed this and Dil Bahadur Ramtel, an eleven-year-old lower caste boy, was killed. He was in the fourth grade when he died and became the first martyr of the People's War.

'The success of our actions during and right after the Initiation gave a boost to the Party and masses in this district. This First Plan of the Party helped us to transform the armed struggle. The lower castes began to join the Party and more members went underground. But some full-timers couldn't stand the repressive situation and left the district. Others also decided to leave the struggle, in all about half of our supporters and activists.

'So the Party decided to raise up the mass movement and we went out to visit people in their houses and organize support for the People's War. We went to the families of those in jail. We went to the homes of full-timers. We talked with the masses about the People's War and they started to think about the Party and how it could help solve our problems. We were able to win back a lot of the people who had dropped away. In this way, we completed the First Plan in the face of many difficulties, but gained much experience.

'At one point we set up a political class with 32 people, and the police raided this meeting and arrested a leading member of the Party. Another Party leader was also arrested, the national chairman of the revolutionary peasant association. Other District Committee members were able to escape, but the enemy captured equipment, money, and documents. The police filed cases against 54 people, and the two leading comrades are still in jail.

'People felt bad that they had failed to save the leading Party members, and the material loss was also a setback. This created a big crisis of political leadership because we had to fill the gap of the leading comrades who had been arrested. One comrade from here was promoted to the Central Committee level and this created a gap on the district level. After a couple of months though, we were able to overcome this crisis.'

After the completion of the First Plan, the CPN (Maoist) outlined the goals of the Second Plan for this district: organize the masses to support guerrilla warfare; to seize arms; and to convert areas of support (still contested by reactionaries) into guerrilla zones and recruit people into the People's Army. The slogan at the time was: 'Develop and promote guerrilla warfare with a plan.' The district leader explains:

'We were able to build up more full-timers, including many women. And we were able to carry out a larger number of actions and more advanced military actions, like the seizure of weapons and raiding police stations. At the same time we had to face the repression of the government's Operation Kilo Sera 2 [a major counterinsurgency police campaign], and we lost many full-time comrades. One woman activist, who was a main leader of the women's organization, was martyred. A district committee comrade and seven guerrillas plus six other activists were also murdered. From one and a half months after the Initiation to October 1998, 14 people were martyred, and there was much grief for these comrades.

'During the Third Plan there was a real zig-zag development of advances and setbacks. During Operation Kilo Sera 2, the police surrounded the district. This was during the monsoon rainy season, the rivers were flooded and the police were guarding all the bridges. So people could not travel outside the district and we lost connection to the Central Committee. The police also destroyed all the revolutionary newspapers coming into this area so people couldn't get them.

'The district was trapped and isolated by this repression for two months. Then the police found out where we had stored most of the weapons we had seized during the Third Plan. For three days, they brought in a helicopter and the guerrillas had to flee the area and leave behind the stored arms. So the government was able to capture these weapons.

'During the Third Plan, the repression was so great it became impossible to do any revolutionary work among the people for three months. Party members lived with the people and continued to talk with them but could not do more active organizing. Then on December 26, we celebrated the birthday of Mao Tsetung by seizing crops from some landlord. About 500 people took part in this action and the guerrillas came out in uniform and defended this action. The police tried to stop the people, but they could not attack such a big crowd.

'After such a long period of repression, we also decided to resist by setting off a bomb outside the quarters of the chief district officer, who controls all the government offices in the district. In another incident, we were able to ambush the police, kill two of them, and capture arms, with no guerrillas injured.

'This district is quite close to the capital and it is easy for government forces to come into this area. So we are trying to build up regions farther north in the hills away from the highways.

'During the Fourth Plan a supporter of UML, who was the secretary of the local Village Development Committee, was annihilated. He had been an informer and had given information to the police that led to the incident in Kerabari where five comrades were killed. In five months we lost 19 full-time comrades who were martyred. Two were district party committee members, six were platoon members, eight were squad members, and three were security forces. In addition, nine people have disappeared, five women and four men. And more than 50 people are still in jail.'

* * *

Late in the evening, we leave for another shelter. There has been too much activity at this location and it is dangerous for us to stay any longer. By the time we hit the trail, the moon is long gone and we have to walk in the complete dark. A woman guerrilla leads the way and I follow right behind. After a while, my eyes get used to the dark and I can see well enough to make out the path for about a two-stride length. I focus on catching the motion of the feet of the woman right in front of me – her green canvas shoes move in a steady rhythm even though the trail is unpredictable and I try to follow in her footsteps.

Close to midnight, we arrive at the house of a very poor family and climb to the second floor, which is just a dirt floor with a straw roof, but no walls. I look out, far away to a mountainside across from us and see huge fires burning. There has been no rain for months and the countryside is as dry as a tinderbox. Fires like this are burning in several areas in the countryside; many peasants have lost their homes and some people have died.

During the day I had seen the snaking lines of smoke which left a heavy haze throughout the area. But now night has taken charge, revealing the extent of damage and danger. I can really see how big the fires are and how many there are. Big, jagged lines of flames are progressing across the mountain ridges, consuming the terrain. The landscape is hidden, completely covered by the moonless dark. So it looks like some kind of crazy, orange lightning is zipping through the night sky. We are miles and miles away. But loud popping and crackling reaches our ears, as if we are standing next to a stove of frying bacon and popping corn. I wake up several times and notice that the fires actually seem to be dying down a little. But after the sun comes up, winds start fanning the flames with renewed meanness. The infernos begin to build once again and the roar becomes even louder. Many people believe the police start these fires in order to counter the People's War.

We set out again as the sun inches toward the horizon, replacing darkness with a faint gray. We get to a small group of simple houses and the extreme poverty here is immediately apparent. As soon as we arrive the villagers put some mats down for us to sit on and everyone gathers round to talk. At first it is mainly men and children but a little later, a number of women emerge from their chores and join the group.

This is a village that supports the People's War, providing shelter and food for the guerrillas. The first thing they tell me is that when the police come and interrogate them no one gives away any information or secrets. They are very proud of this. Then they take turns telling me about their lives – the daily, hard routine that's necessary for survival. A 50-year-old man tells me:

> 'We work more than 18 hours a day and still can't grow enough food to feed our family for a whole year. I borrow money from the moneylender at high interest, 36 percent interest for one year. We get loans to be able to eat, celebrate festivals, or arrange a marriage celebration. I went to the city to work as a laborer and brought back home only enough money to pay the interest. One of the conditions of the loan was to work one to two days a month for the landlord. Plus I had to give him milk from the goats. The landlords are happy, getting money and doing no work. They live in the city and other areas. I only own 0.1 hectare of land. The People's War is for our class against the bourgeoisie and reactionaries and when we win we will be able to have a better life. So I am looking to the bright future of the People's War in Nepal.'

I ask a 27-year-old woman to tell me about her day and at first she says she has nothing to say. But then once she starts talking the words tumble out. I can hear in her voice how it is liberating for her to even speak out about how hard her life is. She says:

> 'I wake up at 5:00 am, prepare a simple soup for the family, get grass for the goat – which takes five hours to get, about three kilometers away – and return about 12 noon. Then I have to clean the pots and dishes. I prepare food and eat and then take the goats and cows to graze them. We have five goats, but we just take care of them for a landlord or someone else. We have three cows but own only two. I take them to the same area far away to graze. I also gather roots in the forest [Githa], which are boiled and put in salt and ash which neutralize the bitterness. It is five o'clock by the time I get back from grazing the animals. Then I have to prepare another meal. I also have to gather firewood from the forest. I finally go to sleep at 9:00 pm. I have three children, one son and two daughters. I am not educated so it is hard to express my grief. We give shelter to the People's Army and through this find out about the struggle. I listen to the discussions that go on when they come.'

A 23-year-old woman speaks next. She says:

> 'I have a one-month-old son and two other children, 7 and 3 years old. I didn't go to school because my father and mother didn't send me. If I went to school there would have been no one to tend the animals

and do other chores so I didn't get to go. I got married when I was 15 years old.

'Women here usually get married between 15 and 22 years old and my people of the Praja nationality don't do arranged marriages, so this was a love marriage. I didn't go to live in my husband's house. I live with my father and mother. My husband is 24 years old and used to work as a laborer. But now he's been working in a restaurant for the last three years near the roadside. He works a one-hour walk away and comes home once a week.'

Like this woman's husband, many other men here have to leave the village for many months to find work. A 45-year-old man tells me: 'I have six in my family, four kids. I have about 0.1 hectare of land and can grow food enough for only four months. The rest of the time I work on the roadside as a coolie. Sometimes the landlord will give us food out of charity. When we work on the road we get 60 rupees (less than $1) for a day with no food. This is not enough to save any money.'

When I ask him what he thinks about the People's War he says, 'When we talk with the People's Army and the Party we feel this is the way forward. And if the revolution is successful the future will be good.'

By now it is almost 9:00 am. I stop to photograph some of the villagers and then it is time to go.[1] The walk down the mountain goes by quickly and before I know it, we are near the main highway. It is dangerous for the guerrillas to go any further, so we say goodbye and they give me a *lal salaam*, the red salute.

9
Preparing the Ground in the West

Going to the Western Region is the most difficult part of my trip to arrange. This is where the People's War is the strongest and so especially now, before the elections, it's very intense in this part of the country. There are stepped-up actions by the guerrillas and more police are being sent in to try and suffocate the region with a heavy blanket of repression. The government is so worried that revolutionary activity will disrupt their election they're doing the voting in two phases, on May 3 and May 17, so they can deploy enough security forces in each district.

All the 'sensitive areas,' where the People's War is strong, are having the elections in the first phase. For this reason, at first, it seemed too risky for me to go to the west before the elections. But leaders of the CPN (Maoist) there send a message saying they think they can make the necessary arrangements.

Back in the US, I had read whatever news I could find about encounters between the Nepalese police and Maoist guerrillas and it had become clear, even from scattered news items, that the strength of the People's War was centered in the Western Region, especially in the districts of Rolpa and Rukum. These were the districts I was headed for, so I was both excited and nervous.

After a 16½-hour bus ride, Pravat, my translator, and I arrive in the Sallyan district. We have a four-hour walk ahead of us to a village in Rolpa where we will spend our first night in the west. The trek is very pleasant in the early morning coolness and now we are only hours away from entering the heart of this People's War.

A little after 9:00 am we cross the border into Rolpa. I look over at Pravat, who is from another part of the country and has never been to the west and I see that he is beaming with excitement. I say to him, 'I have read a lot about Rolpa and now I am here!' And he says, 'Me too! I am thinking the same thing. In my area we are always looking to what's happening in Rolpa and now my heart is filled with joy to be coming here!'

My first morning in Rolpa is the beginning of a two-day meeting with a member of the Party's Central Committee and its Political Bureau. He is in charge of work in the Rolpa/Rukum/Jajarkot districts and begins by giving me some background about the political and economic situation in Nepal:

> 'The Western Region, economically and socially, has a feudal character. Growth of capitalism is very little and slow. There are many social contradictions and the main problem is feudalism. Economically and socially, feudalism has dominated in this area in forms of exploitation. The main contradiction is between the people and usurers. Usurers are also social and political and religious leaders in society. Before the Initiation everything was centralized on usurer power. This contradiction has only changed a little, but not fundamentally.'

In this part of the country, like much of Nepal, there aren't a lot of big landlords and most peasants own the land they toil on. But more often than not, they cannot grow enough food to feed their families. They constantly face the risk of being pushed from poverty into complete destitution by a corrupt official who finds some way to steal their land or a moneylender, a usurer, who charges exorbitant interest. Because the peasants barely survive, ever on the brink, they frequently have to borrow. A farmer may need a small loan if bad weather has destroyed a crop, for new seed to plant, when a son or daughter gets married, or when someone in the family needs medicine. A money-grubbing usurer may charge the peasant anything from 60 percent to 120 percent interest per year on a loan, which means the farmer often can never repay the loan.

These usurers play a key role in enforcing the oppressive feudal property and social relations in the countryside. They are usually 'big shots' in the village and may have connections to government officials and the police. In general, they lord it over the villagers and abuse them in many different ways. The people commonly call the usurers 'liars and cheaters' because they so dishonestly rip off the people.

While traveling in Nepal, I constantly hear stories of how these 'liars and cheaters' oppress the peasants. For example, a villager will take out a loan for 1,000 rupees and the usurer will write down on the loan papers that the peasant borrowed 3,000 rupees. When a usurer doesn't get his money he may confiscate a farmer's land or take one of his animals. Some peasants end up having to work for nothing for a usurer as a condition of the loan. Or a farmer may end up working like an indentured servant to a moneylender because he can't pay back a loan. For all these reasons, usurers in the countryside are a big target of the People's War.

The CC member describes how the usurers in the countryside went from being oppressors under the old panchayat system to oppressors under the present regime:

> 'After the multi-party system was instituted in 1991, usurers plus the government local powers (the police and government authorities) joined forces to oppress the people. The usurers had been *panchas* – part of the panchayat, monarchical system. Then, they turned to the ruling Nepali Congress (NC) and a few others became members of other reactionary parties, like the Rashtriya Prajatantra Party (RPP) and the Communist Party of Nepal (Unified Marxist-Leninist). The masses could see that there was really no political change. Most of the students, teachers, and others were dissatisfied with the multi-party system and the intellectuals also saw that there was no change. The Party also said this and pointed to the need for People's War and this helped the younger generation and others to see the truth and laid the basis for them to rebel.
>
> 'The youth were looking for a revolutionary party, looking to see which one would lead the way to solve the people's problems. And the Party analyzed this situation and gave a revolutionary program to the young generation. The common interests were joined between the Party and the rebellious generation. This was the case especially in Rukum, Rolpa, and Jajarkot. With the change in the government to a multi-party system, the Party was able to stay on the revolutionary path and didn't divert to revisionism or reactionary thinking.
>
> 'In the period 1990–91 [when the government changed to a multi-party system] up to the Initiation of the People's War, the contradiction grew between usurers and the Party and there were many violent actions against usurers. For example, in Rolpa and Rukum there was the case of this one man who was a *pancha* and then became a member of the Nepali Congress. People said he was bad and the Party said he was bad and so the Party led an action to capture this usurer. The people painted his face, put his shoes around his neck (a "shoe garland") to humiliate him, and gave him a beating. This action exposed the crimes of this usurer and it showed how liars, cheaters, and usurers exploit the people economically, politically, and socially. After this action many masses were happy and came around to support the Party and the revolutionary way. And at the same time the Party was organizing in the mass organizations.

'The usurers have no military but they are the enemy at the local level. After this action the usurers went to the police to ask for security. They told the government they would not live in the villages without security because it was too dangerous for them. They asked the government to punish the Party for the "crimes" against the usurers. The usurers were saved by the NC and the NC party's government – and also to some extent by UML and RPP who supported and unleashed all kinds of repression and filed many false cases against people. These parties also worked as spies against the people.

'In order to counter this situation the Party decided to intensify the class struggle, especially in the rural areas. When cases were registered against many young comrades we decided not to go to court for these cases and instead people went underground. We educated people about the role of usurers, the government, and the different reactionary and revisionist parties. Mass organizations were mobilized to counter and expose the spies and police. The orientation was toward doing illegal work. This was part of the preparation for starting the People's War. The situation in this zone, especially in Rolpa, was a major factor in the decision to launch the People's War in February 1996.

'The Party saw the work in this area as a model, a lesson for how other areas could be developed. Before the Initiation none of the actions were armed, but enemies were physically beaten. And there was one instance during this period of using a bomb to attack a rice mill owned by a usurer. All this work laid the preparation for the Initiation. People could see the successes in these districts and the path and possibility to wage the class struggle and armed struggle. The decision was made for a one-year period of preparation for the Initiation.'

The government portrays this revolution as a war that is intimidating and coercing people. The media reports that the guerrillas extort villagers and force them to provide shelter. It makes it seem like the People's Army is just a small band of 'terrorists' who have no support among the peasants. Being in this stronghold of the People's War provides an opportunity to hear a different side to this story.

We spend many, many hours traveling up and down remote mountains. It may not be like this in other parts of the country. But this is an area where the Maoists have established a lot of support and other political forces have been, to a large extent, driven out. Villagers come out to greet the guerrillas, and peasants along the road give us information – like whether or not there are police or other enemies in the area. When we run out of water villagers share what they have hauled up the mountainside. And late at night, or sometimes a few hours before dawn, when we knock on a door and a sleepy peasant answers and sees members of the People's Army, we are let in and given a meal and a place to rest. This kind of support has been built up through a number of mass campaigns carried out by the Maoists in these western districts.

In the year of preparation before the Initiation, the campaigns carried out in the main areas in the west were thought out, well planned and flexibly executed. I am told that the principles applied during these campaigns were: relying on the people, using strength to overcome weakness, combining centralization with decentralization, and giving people a strategic vision of seizing power through armed struggle and moving on to build a new society. The CC member recounts how these campaigns laid the basis for starting the armed struggle:

> 'The first campaign, to centralize forces, was accomplished in the Western Region in one-and-a-half months. Party members, cultural leaders, and leaders of mass organizations from other districts in the region came to learn from what had been done in this main zone. Then, after two or three months, these forces were decentralized back to the different areas and other Party leaders were also sent into these areas to help develop the work. The whole work of the Party, organizationally, politically, and ideologically, in the region, was strengthened through this whole process.
>
> 'The second campaign was a "Friendship Campaign" carried out only in Rolpa and Rukum. Some selected comrades and some cultural teams from Rolpa went to Rukum. And selected comrades and cultural teams from Rukum went to Rolpa. This was a one-month program. Production/construction teams were also involved in this campaign.
>
> 'The purpose of this program was to rouse the masses and heighten political consciousness. These teams of leaders worked with the masses building roads and bridges and farming, involving up to 1,000 people. Such programs were also launched in other districts under the Western Command. This campaign helped to educate the people politically, organize them, and develop economic production and growth. It also included building schools and toilets and water taps. This was called a "People-Solving Campaign." The exchanges between Rolpa and Rukum enabled people to exchange experiences and show solidarity.
>
> 'The total campaign of Centralization/Friendship/People-Solving was called the Sija Movement, after the two big mountains, Sisne Himal in

Rukum and Jaljala Mountain in Rolpa. The Sija campaign lasted (on and off) one year.

'Right before the Initiation, from October to November 1995, the government launched a big campaign of repression, especially in Rolpa and secondarily in Rukum, called Operation Romeo. This was the government's response to the advances of the Sija Movement.

'In the beginning of October 1995, the local Party in one village in Rolpa organized a cultural program. The NC, RPP, and UML organized a group of about 50 to 60 people to go to the program and attack it. There were about 1,000 people at the program and fighting broke out between the reactionaries and us. At the time, the government was trying to find out where and when the People's War was going to be launched and they thought that maybe this program was signaling the start of the People's War. So they launched the repressive Operation Romeo.

'The reactionaries, who were injured in the fight, went crying to the government and demanded that they do something. Meanwhile, we politically exposed the actions of the government throughout the country. The government arrested nearly 1,000 people during this operation and they tortured people, raped women, and raided and looted houses. At this time, the government didn't go after leaders of legal organizations, but went after activists and sympathizers. The government eventually realized that this was not really the start of the People's War and was exposed for all the crimes it carried out during Operation Romeo. After this, the government was forced to temporarily retreat. Like Mao said, they picked up a rock only to drop it on their own feet. This was a major development leading up to the Initiation of the People's War.

'During this time, internally, the party was prepared to launch the People's War. Operation Romeo was the objective situation that came together with this subjective situation, leading up to February 13, 1996. The People's War would have been launched without Operation Romeo. But the government campaign of repression presented good conditions and an opportunity. At the end of the repression we held a mass meeting in Libang, the Rolpa district headquarters. Some 10,000 to 15,000 people attended and the Party announced that the People's War would start very soon. About 20 to 25 similar meetings were also held throughout the country in December.'

10
Learning Warfare by Waging Warfare in the West

My second night in the Western Region delivers me a good night's rest. My body seems to know it must prepare for a period of intensity. For many weeks now, I will be traveling with the People's Army – on foot, up and down exhausting terrain, much of the time in the dark. We are now in Rolpa where there have been many recent actions by the guerrillas, and the people I am with warn me that we must travel carefully because the police will be looking for revenge.

The government is flooding this part of the country with cops because of the elections and so entering and leaving this region is even more dangerous than usual. But half of the more risky part of this trip – getting into this guerrilla zone – is behind me. Now, while there will still be plenty of danger, people tell me that the police are more afraid to come into these districts and many of the 'reactionaries and snitches' have moved out.

I wake up early and get ready for a second day of discussion with the Central Committee member. I start a morning routine that will become familiar in the weeks ahead. The roosters rouse me as the sun is just beginning to paint the dark sky with misty light. At this hour, it feels more like night than morning, but the village is already bustling with activity. Fires have long been stoked and pots of hot water have come to the boil. Someone brings me some hot milk-tea, prompting me to sit up, and the warmth of the cup feels good in the lingering morning chill.

As I let the tea wake me I think back on what I learned the day before about how the People's War started in this area. The thing that strikes me the most from what I have heard so far is how in the year before the Maoists started actual armed fighting there was a lot of effort to lead different kinds of struggle and build support among the peasants.

Before coming to Nepal my understanding of life in a poor Third World country came from books, TV, and talking with people. But now, traveling through Nepal's countryside, I see with my own eyes how peasants are oppressed by foreign domination and semi-feudalism.

Everywhere I go, evidence jumps out, showing how economic and social life here is dependent on, retarded, and distorted by foreign powers – from the 'made in India' buses we travel in, to the US-controlled aid organizations, to the predominance of Western and Indian music and movies.

We will now be in one of the remotest and poorest areas of Nepal and I'm anxious to learn more about what life is like for these peasants who are always on the edge of life-threatening hunger and poverty. Daily life here is dictated by the rotation of the seasons and the ever-present uncertainty as to whether the earth will be kind enough to hand over enough food for growing children. The weather can be friend or foe. Drought can herald disaster; too much rain can destroy a precious crop. But while nature is a constant adversary, this is not the peasants' chief enemy. The main obstacle preventing a better life here is the fact that Nepal remains a semi-feudal, semi-colonial country dominated by foreign powers.

Ninety percent of Nepal's population live in the countryside and *land* is a central issue in the peasants' fight against oppression. And when it comes to land in Nepal, there is great inequality. Landless peasants work on other people's land. In the Terai, many peasants work on plots owned by big landlords. In the mountainous areas, which make up most of the rest of the country, poor peasants mainly own the land they work on, but struggle to feed their families on very small and inadequate parcels of land and are always in danger of losing their property to heartless moneylenders and other scoundrels. A lot of the hardship stories I hear center on 'not enough land' and 'owing too much to moneylenders.'

The CPN (Maoist) believes in the concept of a 'new democratic revolution' that was developed by Mao Tsetung in the Chinese Revolution. The premise behind such a revolution is that to achieve liberation in a country like Nepal, the people must overthrow the bureaucrat-capitalist class and state system, which are dependent on and serve imperialism; they must uproot semi-feudalism in the countryside; and they must kick out foreign capitalism. The peasant demand of 'land to the tiller' is seen as a central part of getting rid of inequalities in the countryside. And developing more communal forms of owning and working the land is viewed as an essential part of breaking free of foreign domination and preparing to go directly to a second stage of socialism.

After breakfast, I start the second day of discussion with the CC member in charge of the Rolpa, Rukum, and Jajarkot districts. Today he will give me an overview of how the People's War started in this area and how it has advanced in the last three years. He starts off with a story about one of the first actions after the armed struggle started:

'During the Initiation stage we carried out four different kinds of policies/ actions: guerrilla actions, sabotage, propaganda, and annihilation. We started by raiding two police posts. One was Holeri in Rolpa. The other was Radi in Rukum. Both posts had about seven to nine police and they were taken at the same time at night. In Holeri the fighting lasted two hours and even after this the police didn't come out and we ran out of ammunition so we had to retreat. In Radi the police surrendered right away and we seized their papers and burned them. Two days later, in Jajarkot, we attacked a big moneylender, Dip Bahadur Singha, a former assistant minister in the old reactionary Panchayat government.

'Dip Bahadur Singha had a lot of money, enough coins to fill seven big water pots, which he buried in a trench. Another seven pots of coins were buried under the floor and some covered by a thick mud wall. He had many brothers and this was their common property. He had told them he was keeping the money for them, but his intention was to keep it all for himself. Singha had been targeted by the People's War and one of his brothers came to the Party and gave information which helped us to carry out the action. When people went to the house Singha wasn't at home, only his old mother was there. The people dug out all the pots and they were all empty. But some coins found in another part of the house were confiscated. Some of Singha's brothers still support the Party and want to get revenge on Singha who now lives in Kathmandu.'

The CC member says this story is typical of how the People's War has targeted moneylenders who abuse the peasants and drive many of them into conditions of gnawing hunger, tattered clothes, and endless debt. He continues:

'I'll give you one example of how Singha earned all his money by exploiting the peasants. He was a big man in the village. One day he went and saw that there was good cultivation of millet on one farmer's small piece of land. He ordered the peasant to put aside for him two kilos of the millet for seed. The peasant kept the millet for Singha but then didn't have any millet seed for his own needs, so he asked Singha if he could use the two kilograms for a new crop. Singha said OK and the peasant sowed the seed. Then at harvest time Singha claimed that the cultivation was his. The total value of the harvest was 8,000 rupees and Singha demanded the peasant pay this amount. The peasant didn't have any money so this usurer took the peasant's milk buffalo. When our forces captured Singha's house there were a lot of false bonds there worth 800,000 rupees. These were all taken and burned. The people of the two or three villages oppressed by Singha then saw the Party as leading the struggle against their oppressors and they are now strong supporters of the revolution.'

From the beginning, the Central Committee of the CPN (Maoist) developed overall plans for the Initiation of armed struggle and then subsequent stages to escalate, spread and deepen the People's War. The CC member explains how these plans were implemented in these key districts:

'After the first days of Initiation, for three weeks, there were many sabotage and propaganda actions. After this we justified these actions by doing propaganda among the people. This movement continued for two or three months, with no further actions. This was the First Plan, the Initiation. The Second Plan started six months after the Initiation. At this time a squad was formed – there were only 'fighting groups' in the initiation phase. In three districts, 32 squads (of seven to nine people) were formed in the Second Plan. In the beginning these squads were more quantitative than qualitative (in terms of military training) and armed only with homemade guns. After the squads were formed the main goal was to convert the zone here into a guerrilla zone, where the police are confronted and in danger from all kinds of actions. Sometimes the squads had to retreat when the police came into a village where the squad lived. But when the police left the squads returned and the village would be back in the political and military control of the People's Army.

'In the Second Plan there were many ambushes of police, going from smaller to larger actions, and also many raids of police posts and mining of roads where police were traveling. Some were successful, some not, due to lack of experience. We were learning the art of warfare through waging warfare. Our success was that many police were killed in the Second Plan (20 to 30) with no fatalities among the people. A lot of sabotage occurred such as capturing false bonds from moneylenders and destroying agricultural banks and seizing their bonds too. There were 30 to 40 annihilations of spies, usurers, liars, and rapists.

'The government reacted with vigorous and random repression, with arrests, murders, rapes, looting, burning of people's houses, etc. The government killed more than 150 people in the one-year of the Second Plan and filed many false cases against people. At the time we exposed the government's crimes and educated the people politically. Human rights activists came and also exposed what the government was doing. Through this political exposure, the government was forced to temporarily back off.

'In this period we realized our military skills were not adequate and needed to be improved. We also needed to improve our working style in terms of working underground, to be more in line with war conditions. Our old style of working was still too much from before the Initiation of the armed struggle and this made it easier for the government to capture and kill people. The last part of the Second Plan was the boycott of the local elections, and in this zone 50 VDCs had no representative and many others were only partially represented.

'The Third Plan started 18 months after the Initiation and the target of this plan was to improve our military strength and power but we realized that the guerrilla zone had to be advanced. We managed a military training program and collected guns and ammunition from three sources – production, buying, and capturing. In the Third Plan our military power was improved and more military actions were launched. Mass movements/organizations gained strength and did political work among the people. And we also did construction of roads, bridges, channels, martyr's monuments, production (collective cultivation), and helping martyrs' families. There are only a few big landlords in this zone but some lands were seized after landlords left and are now being cultivated by the peasants.

'In this zone, the work all basically went underground, including the mass organizations. But if police are not in the area there are still open meetings, demonstrations, cultural events, and so on. On the second anniversary of the Initiation, the mass mobilization was very good. There was a one-month campaign to celebrate two years of the People's War. The time was divided into a first week of propaganda; a second week of seminars, demonstrations, and discussion; a third week of mass mobilization for production and construction; and a fourth week of evaluation of the past year of the war. There was a big festival to celebrate. Presents were given to martyrs' families and gifts of special food were sent to regional and district headquarters. In this anniversary there were also many successful military actions and some rifles were captured.'

In the first two years of the People's War, the police were hit very hard, especially in Rolpa, Rukum, and Jajarkot. The government issued proclamations about how the Maoist guerrillas were small and isolated and would be easily defeated. But meanwhile, they were busy trying to figure out how to put a stop to a revolution that was clearly spreading and gaining popularity. From February to June 1998, the police were inactive, afraid of confrontations with the People's Army. But this proved to be no more than the lull before a new counter-revolutionary storm.

The CC member tells me that after only two years of armed struggle, there were areas where the Party, the army, and the masses were able to start setting up and exercising new, nascent forms of 'people's power.' Then the government launched a major campaign against the guerrillas:

> 'In the villages, the People's Army was able to walk around openly and freely with rifles and uniforms. Peasants registered the buying and selling of land with the Party. Party cadres supervised the village schools. There was a civil court established with a '3-in-1 committee' made up of the Party, People's Army, and United Front (members of mass organizations) to adjudicate and settle disputes. At this time, the first level of enemies – the biggest reactionaries and oppressors – were driven out to district headquarters. A second level of enemies – supporters of the main enemies – was neutralized. They surrendered by making a commitment to not do anything against the Party, the People's War, or the People's Army, and gave us money.
>
> 'This period when the police did not come into the area lasted about six months. Then the government launched the repressive Operation Kilo Sera 2. The government attacked every sector of the movement, arresting activists, villagers, and sympathizers. This was not random but very well planned and they used many spies to target people. The ruling party, the Nepali Congress, and the Rashtriya Prajatantra Party participated openly in this repression while the UML did so in a more covert way. There were many massacres and a lot of responsible comrades, mass leaders, Party leaders, regional and district committee members, People's Army leaders, and sympathizers were killed. Kilo Sera 2 lasted two months, from mid-June to August 1998, and about 200 people were killed in the western region – 15 in Rolpa, 20 in Rukum, over 50 in Jajarkot, and the rest in areas encircling this zone.
>
> 'Before the Third Plan, our forces were concentrated in this zone. In the Third Plan, our forces were decentralized to other zones and many

programs were launched throughout the entire region. The government was afraid of the People's War expanding throughout the whole Western Region so they centralized their forces to try and contain the revolution. By the Third Plan we had over 40 squads and our focus had been on qualitative development. In the Third Plan some squads were sent to other regions and the squads also became larger.

'The Kilo Sera 2 repression made it necessary to go to the Fourth Plan of moving toward establishing base areas. Base areas are especially needed to exercise people's power. And the Central Committee decided to launch the Fourth Plan at the end of 1998.

'At the beginning, there was a campaign to announce the Fourth Plan nationally and secondly, to fight Kilo Sera 2. The character of the campaign was both political and military – the slogan was, "Go Forward to Establish Base Areas." This became a force against Kilo Sera 2 and pushed the enemy back and raised the political consciousness of the people. A plan was made to promote the military viewpoint and many squads came together to form platoons. Now there are several platoons which function as our main forces. The squads are our secondary forces and the militias are the third forces. In each platoon there is one commander, one vice-commander and three section commanders.

'Around this time the government announced special elections because the past ones failed. Many people didn't vote but the government cast many false votes. Even where there were no candidates in the villages and district headquarters, the government announced elections. VDC chairmen and other district members gave their resignation, after this was suggested by the People's War. If they didn't resign they were told they would be forced out. Now most VDCs are not officially represented and the few remaining VDC chairmen have left their village to live in the district headquarters.'

When platoons were formed in the Western Region the Party made plans to escalate the armed struggle and carry out bigger military actions. The CC member ends this session by recounting the most recent military successes in carrying out the Fourth Plan:

'There have been many advanced military actions in the Fourth Plan, which started on October 27, 1998. In Sallyan, the Jhimpe Communications Tower was raided and eight rifles, one pistol, and 600 bullets were seized. In Jhelneta, police on patrol were attacked by the People's Army. Four rifles were seized, the police surrendered immediately, and two police were injured. Some bullets were also seized. In Dolpa, a group of patrolling police were hit by land mines and one rifle was seized, two damaged completely, and five police killed, with no harm to any revolutionaries. This Dolpa action was carried out by Party leaders and members, not the People's Army.

'In Kalikot district a police post was raided by a squad which had only muzzle rifles, and three rifles were seized and two police were killed. The squad fired on the police and then let them run out of ammunition so they had to surrender. In Jelwang our platoon raided a police post, captured and killed the sentry, and seized his rifle. The post was attacked and there was fighting for an hour before the police ran away. The post was blasted and completely destroyed on April 2, 1999. One squad member was killed. In Dang, Chiraghat, the police post was raided by a centralized platoon. Six rifles, one pistol, and 300 bullets were captured and seven police were killed. One squad member was martyred in this action, which was done in celebration of the third anniversary of the Initiation of the People's War.'

* * *

After we finish this discussion others come into the room to talk about plans for my trip through the Western Region. They are anxious for me to see as much as possible, but my time is limited so there are decisions to be made about how much territory to cover. Of utmost concern is security since we do not want to encounter the police or people hostile to the Maoists, who might snitch. The terrain is very rough. I will find these mountains full of challenges that never quit and we will have to do a lot of our travel in the dark without flashlights in order to remain undetected by the police. At one point the CC member looks up from the map he's using to plot our course and asks me, 'Do you think you can walk for 14 hours up and down the mountains in the dark?'

I take a deep breath and don't answer right away. I'm determined to persevere and face a lot of hardship. But I don't want to make any promises I can't keep. After a few moments I say, 'My spirit says I can do it, but I'm not sure my body will agree.' Everyone laughs as they continue drawing up our itinerary. I look down over their shoulders at the map, wondering what lies hidden in this rugged territory, excited with anticipation.

1 Revolutionary cultural program in Rolpa.

2 Revolutionary cultural program in Rolpa.

3 Women in eastern Nepal.

4 People's Army guerrillas in Rolpa.

5 Eastern Nepal. In the distance, near the tree, a People's Army platoon building a memorial for revolutionary martyrs.

6 Relatives of Kami Buda, a member of the Communist Party, killed by the government in 1955. Daughter, Moti Kali Pun, was three years old when her father was killed. Aas Mali is Kami Buda's sister, Rukum District.

7 Wife and child of a guerrilla killed in the People's War, eastern Nepal.

8 Mohan Lal B.K. and Pabita B.K., parents of Obi Ram B.K., 23-year-old guerrilla killed in 1998, Rukum District.

9 Man with Maoist revolutionary newspaper, wearing a hat with shrapnel holes, taken from an election official hit by a landmine, Rukum District.

10 Village in eastern Nepal. The slogan on the wall reads: Get Rid of Oppressive Laws. Release Political Prisoners. Eliminate Liars and Conspirators [signed] CPN (Maoist).

11 Revolutionary posters in a village in Rolpa.

12 Women militia members in Rolpa.

13 Villagers greeting the author in Rolpa.

14 Local militia in Rukum.

15 Women guerrillas in the Middle Region.

16 Woman guerrilla in Rukum.

17 People's Army guerrilla exercises in Rukum.

18 Women militia members in Rolpa.

19 Woman guerrilla carrying a handmade grenade, Rukum.

20 Peasant farmers in the Middle Region.

21 Village in eastern Nepal.

22 Daily life in Rolpa.

23 School children in eastern Nepal.

24 Shepherds who provided shelter for guerrillas in Rukum.

25 Children in a village in the Middle Region.

26 Children in a village in the Middle Region.

27 Local militia in Rolpa.

28 Local militia in Rolpa.

29 Village women in the Middle Region.

30 Villagers at a revolutionary program in Rolpa.

31 Woman guerrilla in Rolpa.

Carrying the Story Forward: Revolutionary Policies

In October 2003, the CPN (Maoist) published a document from a meeting of the Party's Politbureau.[1] One section of the document, 'Refinement in the Practice of Military Actions,' discusses and reevaluates some of the Party's policies. It starts by saying that the Nepalese regime no longer has power in the entire rural countryside – that such areas are now under the leadership and influence of the Party – and then goes on to say: 'In such a situation the masses expect the Party to have all the more seriousness and sense of responsibility. On the contrary, the practices of certain forms of our military actions in certain contexts have now been inconsistent with the level of development of the movement, [our] responsibility and the expectations of the masses. This will ultimately give rise to negative consequences.' The document then discusses policies where it says, 'it is necessary to clarify our concept and practice in the context of immediate military actions.' I found this whole discussion a very interesting look into the real-life practice of making revolution – including seriously evaluating policies, determining if they are correct, or are being correctly applied, and then making necessary adjustments. For example:

'ON DESTRUCTION AND CONSTRUCTION'

When I was in Nepal the guerrillas told me how they had raided the Jhimpe communications tower in Rolpa, and I learned they had also destroyed bridges and roads – sometimes to prevent the police from coming into the area to attack them. Throughout 2002 many news reports and editorials appeared attacking the Maoists for actions like these, which targeted the country's infrastructure. A common theme here was that the Maoists once had widespread support, but that peasants were now turning against the guerrillas because they were wrecking the country and causing even more suffering for the people.

In December 2002 Prachanda issued a statement saying the PLA would stop all activities aimed at sabotaging 'physical infrastructures directly linked to the people's welfare.' Some of the thinking behind this new policy is laid out in a section of the October 2003 document titled, 'On Destruction and Construction,' which says: 'Until the old state power is completely destroyed or until the revolution is successful, strategically destruction is the principal aspect. However, tactically and practically in areas and levels where the old state power is destroyed, the construction (aspect) gains priority in those areas and levels.' It then goes on to say that some military actions taken by the PLA have not properly expressed this relationship between destruction and construction – like setting fire to government buildings in areas already under Maoist control, blasting buildings already vacated by the police, and the sabotage of small businesses even when they are willing to follow the new revolutionary policies and rules. And these things 'smack of assuming, destruction in an absolute sense' and 'raise the danger of increasing people's grievances against us and the enemy's capitalization on it.'

'ON ANNIHILATION OF CLASS ENEMIES AND SPIES'

News articles have reported that the Maoists kill people simply because they are members of opposing political parties. But such reports frequently fail to mention that the people targeted have done things that have directly led to the death of Maoists or people simply suspected of being Maoists or Maoist sympathizers.

For example, the first guerrillas I met were later killed after someone from the UML led the police to the house where the squad was staying. The guerrillas retaliated against this person, not because he was a member of an opposing party as such, but because he was directly complicit in the death of seven people.

The October 2003 document explains: 'As per the physical liquidation of class enemies and spies, our Party's policy has been: to practice it on selected ones and to the minimum, by informing the masses and obtaining their consent as far as possible and by not resorting to any ghastly methods.' However, it seems there have been deviations from this policy. The document goes on to say:

> While annihilating somebody if we fail to develop and observe concrete policy on class analysis, the nature of his/her crime, democratic legal process to establish the crime and the method of annihilation, it may have negative consequences. It can't just be dismissed as a baseless charge of the enemy and the opportunists that in the past some of the annihilations have taken place flimsily on the grounds of not giving

> enough donations, not providing shelter and food, having politically opposed our movement, suspicion of being a spy, or having enmity with our local team members.

A later section explains that the party's policy 'On Dealing with Opposition Political Leaders and Cadres' has been to 'give precedence to their political exposure' and that 'any physical action against somebody should be not because of his/her membership of a particular political party but due to his/her crime against the people and the People's War.' The document criticizes some actions that have gone against this policy and says, 'In the new situation we should on the one hand, strive to strictly implement the above policy and, on the other, strive to follow the earlier mentioned democratic legal process in the context of such persons deserving any kind of punishment or physical action.'

'ON ACTION AGAINST ENEMY SOLDIERS AND POLICE FORCES'

The October 2003 document reiterates that the PLA's policy toward enemy soldiers and policemen who surrender is 'to deal honorably with the prisoners of war, to convince them and provide them opportunity either to join us in the fight or to return home.' It then says, 'It is against the principle and practice of the People's War to liquidate someone when encountered alone, or at home on leave, or anywhere on the spot just because he happens to earn a living in the enemy's army or the police force, and ultimately its consequence is to unify rather than disintegrate the enemy.'

The guerrillas are therefore ordered not to attack someone in the RNA or the police who is returning home on leave. Instead such people are to be 'indoctrinated and persuaded to abandon the [enemy's] service.' And the correct policy of the PLA 'is to enlighten the family members of those serving in the enemy's army and police force, to organize them and protect them.'

'ON COLLECTION OF DONATIONS'

In recent years, it has not been unusual to pick up the travel section of a major US newspaper and read an account by trekkers in Nepal who have been asked for a 'donation' from Maoist guerrillas. Most of these tourists recount this experience in an almost amused way, saying they didn't feel threatened at any time and found it interesting to 'meet the Maoists.' One Canadian trekking guide told a BBC reporter that for many of her clients, talking with Maoist rebels was the high point of their visit and said that she had 'pleasant talks with them and they were quite polite with me.'

The Nepalese government has claimed that the Maoists extort 'donations' from peasants, saying that the guerrillas basically coerce people and punish those who refuse to contribute. This was not my experience traveling in the guerrilla zones where the Maoists clearly had a lot of political and practical support. And it is hard to imagine that the People's War could have advanced as far as it has without real support from the peasants.

At the same time, it does seem like there have been problems with the 'collection of donations.' The October 2003 document says:

> The enemy has been making a big issue of our method of collecting donations to spread illusion [false impressions] amongst the middle classes. To deprive the enemy of this weapon and to prevent the middle classes from getting misled and terrorized, it has been necessary to make our donations policy more systematic and refined. In the past, while collecting donations there have been some lapses in making adequate class analysis of the related persons, organizations, etc, in studying their background and in dealing with them accordingly. As a result a lot of anarchy has been seen in the field of financial collections. Hence in the future this anarchy should be ended and to make the financial collection process more systematic concrete criteria for collecting donations from the people on a class basis should be fixed and, on the other side, a system of punishing and taxing the enemy should be developed.

11
People's Power in Rolpa

After a few days of discussion with the person in charge of this zone, we are ready to start our trek, north and east, into Rolpa. We wake up very early and begin walking at 6:00 am. The mountains in the Western Region loom large and the air is hazy and cool in the morning. When the sun comes up, its rays are blocked and diminished by thick mist. And so, instead of a brilliant sunrise, the day starts with an intense neon, reddish-orange ball rising out of the towering mountains. The security for our travel is very tight and systematic. Some of the squad go up ahead to clear and check the road. Our small group is in the middle and behind us more guerrillas walk vigilant. In this area the guerrillas can walk about openly during the day with their rifles. But we could still encounter police, so we have to be careful.

The paths traverse up and down the mountains and, like a lot of the terrain here, it is very rocky, winding around sheer cliffs. As I negotiate these steep, broken staircases nature has built, I wonder how I'll deal with them after the sun goes down. The squad easily follow these trails very quickly, for many, many miles with little rest. But since they are with me, the going is a lot slower. And although we take ten-minute rests only every two to two and a half hours, this is a lot slower than the guerrillas are used to.

By the time we stop for an evening meal at a peasant's house, we've been traveling for over ten hours. My translator, Pravat, who sees how tired I am, asks whether we can stop sooner for the night. But the person in charge thinks that for security reasons we should press on and reach our destination before the next sunrise.

We hit the trail again and after about a half hour the sun slips behind the mountains. Darkness quickly envelops our convoy. Now I feel like I'm walking with a dark-gray veil over my eyes and it's hard to distinguish potential obstacles on the path. After a while though, my eyes get a little better at seeing in the dark. I also start to use my ears, to take cues from the person walking in front of me. I find that the sound of others' footsteps gives me vision my eyes no longer offer. For the next couple of hours we walk in silence as I concentrate on keeping my weary legs steady and moving.

We stop only twice, for about 15 to 20 minutes. And each time, I immediately sit down, lean against a rock and instantly fall asleep! By 9:00 pm I am mentally and physically exhausted, but the people I am with are very considerate. They anticipate each part of the path they think I will have difficulty with and extend their hands, helping me on slippery rocks across rivers, steadying me on narrow paths covered with a thick bed of pine needles that have become unstable, and pulling me up the steepest parts of the path. There are many times when I stumble or start to lose my balance in the dark and it amazes me how each time, in a split second, there is a hand that grabs my arm or reaches out quickly to make sure I don't fall. Finally, around 9:30 pm someone next to me whispers, 'We are now getting close to our shelter.' I feel a shot of relief. But then he points into the dark night in the distance, up at the mountain's peak and says, 'Our shelter is at the top.' My initial relief quickly melts – the dark, house-shaped form looks so far away and so high up. But I try to just think about how nice it will be to go to sleep when we get there. When we arrive, it is 10:00 pm – we've been traveling for 16 hours.

The next day we arrive in a village where I get a chance to talk with the person 'in charge' of the whole Rolpa district, as well as leaders of mass organizations and families of people killed by the police. I am particularly interested to learn how the Party's Fourth Plan of moving toward establishing base areas is being implemented here and how they are setting up and carrying out new forms of what they call 'people's power.'

In areas where the People's War is strong, the official government structure has lost much of its authority. Many chairmen of the VDCs set up by the government were forced to resign or moved to the district headquarters. This created a power vacuum, which at first the Party wasn't prepared to fill. But then the Maoists developed a plan to create new forms of power and called on people in the villages to reject the official government.

The Party formed 3-in-1 committees made up of people from the Party, the People's Army, and the mass organizations. These new committees, called revolutionary united front committees, have been given authority to judge and settle various disputes among the people – arguments over land, debt disputes, cases where a husband is beating his wife, divorce petitions, etc. I am told that more and more people have taken their cases to be judged before these committees. And the people have also used these new forms of revolutionary power to deliver justice to those considered 'enemies of the people.'

For example, the people will struggle with spies and snitches and warn them to stop helping the police. But if such people persist in activity considered harmful to the revolution, the 3-in-1 committee will decide they should be punished. In some cases, if what these people have done has resulted in the death of someone, they might be killed.

The 'in charge' in Rolpa explains that there is a relationship between strengthening the People's Army and moving toward establishing a base area and exercising people's power. In Rolpa, the Party has been able to form a number of platoons and many squads, which are crucial to protect the area and prevent the police from coming into the villages at will. And as the guerrillas have carried out successful military actions against the police and local reactionaries, this has resulted in large areas that, while not yet base areas, are places where the authority set up by the Party – militarily, economically, and politically – is mainly in command. The revolutionary hold on these areas may still be fragile – if the police decide to raid the villages or launch a major repressive campaign, the People's Army may have to temporarily leave. But these are areas where there is a lot of support for the People's War and it is difficult for the government to regain any political authority or complete military control. And this is a big part of the basis for the People's Army to return and militarily re-establish people's power when it is lost.

At the very beginning of the Fourth Plan the Party set out to establish areas where the police are afraid to go and the people have relative freedom to exercise new forms of power. One of the first actions of the Fourth Plan was in October 1998, when a platoon took part in the sabotage of a police post in the southern part of Rolpa. The post was totally destroyed and police uniforms, papers, and guns were burned. Two policemen were killed. Soon after this, a People's Army platoon was ambushed in Nimri. The guerrillas were preparing for an action when a spy told the police where they were. The police encircled the platoon but the guerrillas broke through the encirclement and killed one policeman and wounded three others. One platoon member was killed but all the others escaped safely. Then, on February 6, 1999, one week before the third anniversary of the Initiation, three members of a People's Army assault team attacked a group of eleven police on patrol in Jelwang. This included a sub-inspector and two police who were seriously injured. These successes by the People's Army, along with other smaller military actions, made the police afraid to even go out on patrol. The 'in charge' in Rolpa tells me:

> 'After these military actions by the People's Army the police enlarged their forces in every post. After the third anniversary police were not able to move about easily in many places. They changed their tactics. Where before they would come into villages and kill randomly, now they targeted the People's Army and Party cadre. Using spies they encircle the houses where people are staying, move in and kill them. Now the police come into the villages undercover to try and spy. If they are patrolling they have another group move to cover the first group. After the raid of Jelwang they closed down some police posts and combined forces. There are now fewer, but larger centralized police posts. But now the police are hardly ever going out on patrol. This is the achievement of the Jelwang raid. This situation shows that the reactionary state power and its support forces in this district are being eliminated. It means there is a good situation to conduct our people's power. It shows the possibility of establishing a base area.'

I also talk to someone who is introduced to me as the chairman of the United People's Front in Rolpa. He is in charge of developing people's power in this district and tells me that after the Party called for establishing united front organizations to exercise new people's power, one of the first things they did was carry out some investigations:

> 'We gathered population data all over the district – an assessment of ages, number of men and women, religion, politics, language, economic factors, etc. On the basis of these data, including numbers of people who can vote, who are over 18 years old, we decided to make committees in every ward. For example, we set up a united front team of five members where there are 100 to 300 voters; seven members where there are 300 to 500; nine where there are 500 to 700; and eleven where there are 700 and over. Because of security and other organizational problems it was impossible to have an actual election at this point so we gathered the people in a mass meeting to select the united front ward committees. At this point we form teams of nominated united front people but we are aiming to make them elected.
>
> 'The basic principle of forming the revolutionary united front is to elect members on a 3-in-1 basis. For us this means 40 percent from the Party, 20 percent from the People's Army and 40 percent from

the masses. The policy from the Central Committee is that when the people's representative assembly is formed it is supposed to elect two committees – a permanent committee and a functional committee. But now we only have the functional committee and this changes according to developments. We hope in the future to form permanent committees. The District Committee is at this point an ad hoc committee, not elected. Through these functional committees the strategic and tactical work is carried out. And there are different departments such as construction, health, cultural, security, judicial, economic and others.'

I am told that the united front led by the Party is now solving many disputes that occur every day among the people. Lawyers used to come to the area, take people's cases, and charge lots of money. But now in much of this district, there are no longer any political cases, or any other kinds of case, being brought to the government's courts. Some who opposed the Maoists used to register cases against the people but even they no longer go to court here because they are either afraid or know it is useless. Now the only cases being brought before official courts here are those filed by the police against people charged with revolutionary activity.

When people are arrested they have to pay a certain amount of bail money to get out of jail. Now many people go to the person who they claim is responsible for the case, like a landlord or other 'reactionary,' and demand that they give the people this amount of money. I am told there have been about 30 cases like this. The United Front leader also says that some cases over land which have been pending in the courts for 10–15 years, like the case against a usurer who took over public land, are now being solved by the people's courts. The land owned by many 'liars and cheaters' who have exploited people has been recaptured by the people's court and is now being farmed by the peasants.

According to the Party, at the area and district level, there have been about 700 cases brought before the people's court. At the village level there have been a couple of thousand cases. Mass meetings of the people's court have been as big as 500 people. Most of the cases in the people's court here are over land. But there are also all kinds of other cases, like disputes over loans and cases around false bonds – certificates of debt that lie about the amount a peasant owes. One example I am told is about a man in the Ghodagaun VDC who made many false bonds, in all worth 700,000 rupees. After he was brought before the people's court, the Party seized all his bonds. The people's court has also taken up cases of women who have been beaten by their husbands.

The 3-in-1 committees settle cases of various crimes and disputes after consulting with different sections of the people. The committee consults with others for two reasons. One is to decide and finalize the case; the second reason is to decide what kind of punishment is appropriate. The maximum punishment so far has been beating and public humiliation. The minimum has been to make the person confess and say they are sorry. In cases where the person is given a beating, the UF leader says:

> 'If the crime was against a woman by a man, then the punishment is carried out by women. If the case is against a woman by a woman, the punishment is carried out by women. If the case is between two men, the punishment is carried out by men and women. If it is a crime against a man by a woman, punishment is carried out by women. In more serious political crimes, like spying and informing (by men or women) punishment can be carried out by the People's Army. Because women are very oppressed in this society the Party has so far not carried out any annihilations of women.'

The authority of the United Front and people's power is also beginning to take over many other administrative and economic functions in the community. The UF leader explains:

> 'In terms of economic questions and the United Front: The usurers have all left the villages, but the peasants still need to borrow money. So to solve this problem we have established a cooperative financial fund. It is handled by the area committee (not at the unit or district level now). The financial fund is built from two sources. One is by collecting people's savings. The second is from fines from the people's court. There are also fees, like for registering land, using forest materials, etc. The fund makes loans to the people in two categories. If people are sick or need to buy food grain there is no interest. Loans for production are charged 10–15 percent interest per year (which is very low). In the past usurers charged the people 60–120 percent interest per year. There is also a martyr memory fund that donates money to martyrs' families (funds for this are collected from among the people, including at cultural programs and from schools). This fund is used to help martyrs' families and is given according to their financial needs. In terms of health we hold medical examinations and distribute medicine to sick people. We have health assistants who do medical check-ups. We have hygiene and

sanitation campaigns and educate people to go to clinics to deal with health problems and combat beliefs in magic and superstition. Through women's organizations and other mass organizations there are some sex education campaigns.

'Forms of people's power are also beginning to develop some cooperative agricultural production. Land taken from usurers is used to collectively grow food for the People's Army and the Party. And some land is distributed to individual peasants. Some people donate and take care of animals, like chickens and goats, for the Party and the People's Army. And peasants will also buy a small cow or buffalo, raise it, sell it, and then give the profit to the Party or the People's Army.

'In terms of construction in this district, we have built: 510 small martyr platforms under trees; 65 bigger monuments to martyrs; 174 wooden bridges; 135 small and big roads; 75 wells and water taps; 115 school and public toilets; and 15 playgrounds. This is done by the masses, Party members, people's power committees, and the People's Army. There is also cooperative construction of houses for families that need homes.'

* * *

We have been talking for many hours now and it is late in the afternoon. For the last half hour the sounds of some sort of commotion have been wafting into the room from outside. I wonder what's going on, but don't want to disrupt our meeting by getting up to look out the window. Then someone comes into the room and informs us that word has been going round the nearby villages that 'the journalist from the United States' has arrived and spontaneously, villagers have been gathering outside to greet me. We decide to take a break so I can meet these people.

We go outside and I see that over 100 people have gathered on a hillside nearby.[1] The crowd is almost all women and children, and, as we walk up the hillside, I can feel their anticipation and intense curiosity. It is late in the day and the last rays of the sun are emanating from low on the horizon, but the many different colors of people's clothes look bright even in the dusky light. When we get to the top of the hill the crowd erupts in applause. And when I take my camera out to capture the moment, everyone, even the smallest child, raises their fists in the air. The whole scene reminds me once again how politicized the people in these war zones are and how much they want the world to know about their struggle.

12

Guns, Drums, and Keyboards

In the morning I wake up to the rhythm of someone hammering on metal. When I go outside I find some squad members squatting in a circle making grenades. One person is cutting and carving small wooden pins; someone else is hammering away, breaking up a piece of heavy metal into small pieces of shrapnel. Another guerrilla is assembling the parts and adjusting the triggering mechanisms. The People's Army has very few modern rifles – most of the guerrillas I have seen are armed with old muskets – so they have to rely on their khukuries, homemade hand grenades, and land mines. It seems like every guerrilla carries at least a khukuri and grenade, even if they don't have a rifle. I am told that the more sophisticated weapons captured from the police go to the platoons first – rifles to the members and pistols to the platoon leaders.

Traveling and living with the guerrillas is giving me a real sense of what their day-to-day life is like. They are constantly on the move, traveling for hours, mostly during the night. The squad gets support from villagers but they are also self-reliant. When we come into a village the squad gathers firewood and fetches water and they try to carry their own rice with them instead of always eating grains provided by the peasants. They seem to be on good terms with villagers, treat them with respect, and don't take advantage of their hospitality. They're always trying to engage people in political discussion about the aims and goals of the People's War and telling them news about what's happening in the war.

Like in other villages we've visited, some of the peasants here have sons and daughters in the People's Army and they treat the squad members as they would their own children. We are in areas where the Maoists have established a lot of support. Many of the peasants here see the guerrillas as protection from the police and they also see the Maoists as people who help them solve problems, give them political education, and organize them to fight against corrupt politicians, moneylenders, and others who oppress them.

The People's Army also organizes people through the use of revolutionary culture. A cultural squad arrived early this morning while I was still asleep and they immediately started practicing. I spend the day meeting with several leaders of the Rolpa District and all afternoon, as we sit on the floor talking on the second floor of the house where we are staying, the sound of drums, keyboards, and young voices drifts through the village and into our window.

I have been anxiously awaiting the cultural program scheduled for 8:00 pm, so I am disappointed that our meeting doesn't end until 10:00 pm. We still haven't eaten our evening meal, but *dal bhat* and curry potatoes quiet our growling stomachs. Then to my surprise I learn that the cultural program has just started. It was delayed until our meeting was over, and the villagers have been waiting patiently for the 'chief guest journalist' to arrive.

We walk out to a large clearing and I see about 500 people sitting on the ground. A straw screen has been set up at the front and two kerosene torches flicker light for the 'stage.' There's a starry sky overhead, but the color of night that surrounds us is inky black. And while the warmth of the day lingers in the air, a cool breeze has started to rule the temperature. I know we are surrounded by terraced fields, trees, and towering mountains. But in this moment, it feels like we're floating in darkness, in a liberated pocket of revolutionary celebration.

The cultural team has ten members – four women and six men, all very young. They put on an energetic show full of singing, dancing, and skits, accompanied by traditional Nepali drums, guitars, and portable keyboards. Songs of varying rhythms and moods tell tales of guerrilla actions and people killed by the police. Dances combine traditional moves and music with new steps and poses to narrate war stories.

The skits move the crowd to laughter as well as tears. A humorous one is about the upcoming elections and people howl as the character of a fat Nepali Congress candidate swaggers around the stage, blustering fake promises to some peasants. The next skit moves the crowd into a harsher reality with a scene between a young guerrilla and his parents. The mother and father are sad because their son is leaving to fight in the People's War and are afraid he will be killed. The son tells them he must go into combat for the people and that they shouldn't worry. I start to hear quiet sniffling behind me and at the end of the skit I look around and see many people crying. For them, the scene is very real and immediate. My translator whispers into my ear, 'Many people here have relatives who have been killed by the police.'

The program is full of themes – of courage, sacrifice, and celebration of victory – that weave in and out of lyrics, dance moves, and skit dialogue. The crowd seems to be having a good time and I really see how these cultural programs build support for the revolution and give people a sense of collective strength.

In between acts, the MC names people in the crowd who have given five or ten rupee donations. After each announcement the crowd claps and cheers. These are very poor peasants and the donations are small, less than 10 cents. But this is another way villagers support the People's Army and some of this money goes to the families of those who have been killed.

Toward the end of the program the team performs an 'opera' – a drama with singing in which a People's Army squad encounters the police and one guerrilla is killed. The songs are full of sadness but also renewed determination. The fallen guerrilla's body is draped in a red cloth with a hammer and sickle and carried off. Again I hear people in the crowd crying softly.

The program isn't over until after 3:30 am, but everyone has stayed to the end.

Eliminating feudal culture and developing revolutionary culture is a big part of the People's War. Cultural teams were an integral part of the party's preparation before the Initiation, and now the People's Army has cultural squads like this one all over Nepal. The spread of revolutionary culture is an important way the Party popularizes the aims of the revolution, and educates, mobilizes, and recruits peasants into the struggle.

The cultural squads travel from village to village putting on programs. And they also work in the fields with the peasants, participate in people's power committees, and carry out armed actions against the police and other enemies of the People's War. They are constantly developing new songs, dances, and skits. I notice they all carry little dog-eared notebooks in which they write down songs and poems and they frequently pull them out to share with each other. As I travel round the country I notice differences in the songs and dances. But I also recognize common themes and some songs that have managed to migrate from notebook to notebook and become standards.

At our next shelter Man Kumari Bista comes to talk with me. Her husband, Masta Bahadur Bista, was killed by the police in 1996, when he was only 23 years old. She tells me, 'He was underground after the Initiation. He was taking shelter somewhere and the police encircled the house. They came in while he was sleeping and shot him in bed. He was a well-known chairman of a district cultural team. He was a very good musician and singer and wrote many revolutionary songs.'

I am reminded of the cultural squad in the east, which soon after I met them, lost seven members in a police attack. I have been told that many other cultural squad members have been murdered by the police and the government has harassed, arrested, and sometimes killed many progressive and revolutionary artists. In one city the police went to a cultural program and arrested 15 people. Thirteen were released after initially being framed for attacking the police. Two others were released but re-arrested, then taken to another district and framed in another case. One well-known writer of literature told me he had been arrested over 100 times.

In Kathmandu I had interviewed Bhaktu Bahadur Shrestha, who is very involved in promoting revolutionary culture. He is president of the National People's Movement Coordination Committee (NPMCC), which was formed 'to struggle for the cause of democracy, nationality and the livelihood of the people, to oppose state terror unleashed by the government and to fight for various causes of the people.' Talking about government censorship of the arts, he told me, 'Progressive and revolutionary Nepali culture is developing and it is in contradiction with bourgeois culture. The government is trying to smash it and has killed democratic dancers, singers, and artists and some of them are very well known. There have been arrests and harassment all over the country, including police raids on performances. The police and government know who the progressive artists are and arrest and kill them. These are the artists who criticize the government, and expose them. Some of these artists say they support revolution in the world and the People's War in Nepal and they are very popular among the people, especially in the countryside.'

I also interviewed a revolutionary intellectual and a well-known short-story writer, critic, and journalist at Tribhuvan University. He also talked about the government's clampdown on revolutionary artists, as well as how the creation of a new culture is an important element in the People's War. He said:

> 'Since the initiation about two dozen cultural activists have been killed by the reactionaries – singers, dancers, writers. There is a trend of revolutionary art in Nepal. Revolutionary literature in Nepal started in 1950 and now it is led by the All Nepal People's Cultural Organization, which openly supports the People's War and is supported by the People's

War. There are two departments in this organization, writing/literature and dance/drama. The writing department publishes a literary magazine *KALAM* ('pen') and I am the editor of that magazine.

'The cultural activists are mainly in the remote areas and the writers are mainly in the city. Most of the revolutionary writers are being terrorized by the government. I was arrested three times. Last November and January and again last week. They came to my house, confiscated all my revolutionary books and writings, and treated me like a criminal. They came in the night-time. They searched my house for four hours. The police who raided were led by the Deputy Inspector of Police. At the same time other writers were arrested, including Sakti Lamsal, who was arrested at the offices of the revolutionary newspaper *Janadesh*. Nearly a dozen writers are now in jail. Many cultural activists have been forced to go underground. We are charged with anti-terrorist laws. We are arrested as criminals, not writers and journalists, and we are treated inhumanely. Many people oppose this kind of state tyranny, even if they have government jobs. And there is a lot of unity around fighting the suppression of revolutionary and progressive artists.

'Revolutionary writing and cultural activities are very popular. We have produced six cassettes of revolutionary songs, not only in Kathmandu but from other areas as well. Our cultural front is guided by Maoist ideology and the ideas in Mao's *Talks at the Yenan Forum*.[1] Our cultural activists are fighting with both pen and gun. They are a real cultural platoon. We have published short stories, novels, and plays. There are a number of revolutionary critics, poets, playwrights, and novelists who are nationally recognized – and terrorized by the government. Our new revolutionary culture is also very rich in dance and singing and this is very popular among people.

'Mao's *Talks at the Yenan Forum* has been introduced into the Nepal revolutionary cultural front since 1950. Most of the revolutionary writers are inspired by this and even some bourgeois writers in Nepal are influenced by it. It is the basic text for us. Gorky, Howard Fast, Lu Xun, Premchand [from India] are also influential among revolutionaries in Nepal. During the 1960s, artists in Nepal were also very inspired by revolutionary writers who became martyrs in the Maoist Naxalbari movement in India.[2] And the Great Proletarian Cultural Revolution in China was a big influence.

'The People's War has given us energy and we have given it energy. The People's War has given the revolutionary writers a new reality – new subjects, materials, themes for cultural works like base areas, guns, and the armed struggle. In our literature we have talked about guns, guerrillas, and war. And now all this is a reality.'

13
Teachers in a School of War

Our entourage – a People's Army squad, my translator Pravat and me, and a couple of others – are heading for another village today and there will be many goodbyes. The cultural team that has been traveling with us is going off in another direction and several other people who have been with us over the last couple of days are returning to their villages.

The travel today is much easier – we only go for about six hours, mainly in daylight. After we reach our shelter we have a hearty meal of *dal bhat*, curry potatoes, and mutton. Then a teacher from the area comes to be interviewed. I have already met many teachers in the countryside who are involved in this revolution. It seems like village teachers frequently end up 'educating' their students in more than reading and writing.

This teacher has a gentle face that contrasts with his strong, stout body. His voice is soft, but his presence fills the room even when he is silent. Like many of those who have been tortured by the police, he is deadly serious when he talks about their brutality and surrenders no smiles as he recounts his story.

He tells me he started working with the Party in his village shortly after the government adopted the parliamentary system in 1991. Then, in 1996, in the first year of the revolution, the police came to his house, arrested him, and dragged him to a police post two hours away. He says the police beat him the whole time and threatened to kill him on the spot. He was thrown in jail and then for a whole week the police tortured him, trying to get him to give information about the Party and its work. He says, 'I never told them anything, even though I thought they might kill me. After a week they continued to interrogate me. But still, I refused to tell them anything. I only told them I was a sympathizer in the village but nothing more. The police charged me with breaking the "peace and security" law and I was finally released after paying a high bail of 10,000 rupees' (10,000 rupees is about $180, a huge sum of money for the majority of people in Nepal where the per capita yearly income is around $210).

When this teacher is done with his story I tell him that he has given me some insight into one of the reasons the government is having a hard time crushing this revolution. Like many others who support the Maoists, his resolute belief in what he is fighting for gave him the strength to defy the police and refuse to give them any information, even in the face of death.

'What I am doing in the People's War is part of the world revolution,' he says. 'I thought that all the people in the United States were rich. But after meeting you, seeing how you have come from so far away to learn of our struggle, and after hearing about the struggle of people in the US, it makes me even stronger in my revolutionary determination to stand up to the enemy. I see you as an actual physical example of proletarian internationalism and this inspires me. In our People's War many people are being martyred and compared to this, what I have done is little. Now I am no longer underground and do legal work openly. But the Party's work encompasses both legal and underground work and in the future, if I have to, I will go underground again. Now the police are watching me and looking for a chance to arrest and torture me again. And I pledge that if that happens again, I will never give away any Party secrets to the enemy.'

* * *

The next day it is still dark outside when we leave at 4:45 am and head north toward the border between Rolpa and Rukum. Ahead of us is a 14-hour trek before we reach our shelter and we have to clamber up and down many huge mountains. The People's Army squad is with us again, divided up, ahead and behind.

At one point we come to a big river where a bridge is being repaired. It had been built by the government but when it fell apart it was never fixed. So now, as with many other things in Rolpa, people's power committees are mobilizing people to deal with the problem. A team of men are in the process of rebuilding the bridge and the semi-rebuilt span only has slats of wood placed far apart, but not fastened. To walk across you have to hold on to the side railing and jump from loose board to loose board – just a slip away from plunging to the water and rocks far below. As we approach the river, I'm getting my courage up, a bit excited about facing this new challenge. But my traveling companions decide the bridge crossing will be too dangerous for me. That, plus the security situation in this spot, which is near a road sometimes patrolled by the police, means we must pass through this area very quickly. The people I am with say we can't afford the risk of my inexperience.

The guerrilla in charge of our travel motions me to follow him. As everyone else quickly heads for the bridge we go a short distance up the river and round a bend. The water here, while not too deep, rushes over jagged rocks. The guerrilla takes off his shoes and socks and rolls up his pants. Thinking we're both going to wade across I do the same, tie my boot shoelaces together and sling them round my neck. We make our way toward the water and I wince as the soles of my feet hit the rocks. When we get to the river's edge my escort stoops down and motions me to climb onto his back. He is not a large man, but he's incredibly strong and he carries me through the rapids in a steady, zig-zag path to the opposite shore.

When we get to the other side we have a short hike up the mountain to rejoin the rest of the group, who have quickly crossed the bridge. As we approach the top I see they are sitting under a shady tree, looking as if they've been lounging around all morning. I'm still a bit out of breath when we walk up and as someone hands me a bottle of water, I quip, 'I know what you guys are up to. I could have made it across the bridge. This was just a ruse so you could all get some extra rest!' Everyone cracks up laughing.

Later in the day, during one of our rest stops, I ask Sanjeevan to tell me the story of how he came to be what others have described as a 'responsible party comrade.' Sanjeevan, who is 29 years old, has traveled with us since we arrived in Rolpa and has been helping me up, down, and round the mountains and seems to always be very concerned about my health and well-being. He has a sparkle in his eyes, a big smile, and a laugh that peaks a high note with excitement. He has been helping to translate for me and since he is only one of a few revolutionaries I've met in Rolpa who can do this, I am curious to find out how and why he learned English.

Sanjeevan, who was a schoolteacher in his village, has been a member of the Party for quite a long time, since around 1991. But his path to becoming a 'full-timer' in the People's War has been full of twists and turns.

'Although my revolutionary spirit and enthusiasm were very strong,' he tells me, 'at first, I was not that clear on the correct line for how to make revolution.' Because he was a schoolteacher, he was in a Party cell with other more educated cadre who, Sanjeevan tells me, had a more intellectual outlook on the Party's work. He says, 'I would ask them if we could meet more often to discuss the Party's line and various documents. But while these comrades talked about revolution, in practice they were doing something else. Sometimes I didn't even get the Party documents to read!'

Some other teachers in the area, who were close to the ruling Nepali Congress Party and the UML, didn't like Sanjeevan's Maoist politics and arranged for him to be transferred to another village, very far away. But as it turned out, according to Sanjeevan, this transfer ended up being a good thing and a turning point. Sanjeevan says in this new village he was under the leadership of 'comrades with a correct line' and he was finally able to discuss, struggle over, and get clearer on the Party's line – and especially what it meant to prepare for and then actually start a people's war.

Some time after this Sanjeevan went to the city to study English literature. When I ask him why he decided to do this he says, 'Because I wanted to read all the Marxist-Leninist-Maoist works which exist in English but aren't translated into Nepali.' While in the city Sanjeevan searched for other intellectuals who might want to discuss how to solve the country's problems of poverty, exploitation, and government corruption. But every group he talked to was not interested in discussing revolutionary politics; or if they did, Sanjeevan says, it was all talk and no action. Then, in February 1996, the People's War was launched and this sent Sanjeevan into an ideological crisis. He tells me:

> 'I used to sit in my room and I'd be reading and studying all day and then I would go to bed at night and toss and turn all night, thinking about how other comrades were in the countryside waging the People's War while I was sitting in a room reading English literature. I would finally fall asleep and dream that I was with the People's Army fighting the reactionaries. But then I would wake up and still be in my small rented room in the city. I would think for hours, in turmoil, wrestling with myself about what to do.
>
> 'Finally I made a decision. Six months after the Initiation I gave away all my books and left the school without giving notice. I went back to my village and told my family I wanted to join the People's War in the countryside. My mother and father were against me doing this. They told me I should just be a sympathizer, not a full-timer. My wife was also against it. But in the end I had to rebel against all of them and I have been a full-timer in Rolpa ever since.'

* * *

After more days of travel we are near the border between Rolpa and Rukum. We reach our shelter late at night and in the morning I meet with a room full of revolutionary teachers who have come from the surrounding area.

After the start of the People's War some teachers in this area immediately went underground while others continued to do open, legal work. People tell me that now it is increasingly hard to work openly because the government goes after anyone they suspect is sympathetic to the revolution. But they tell me that in spite of these difficult conditions many teachers in this area support the People's War. One teacher starts off the discussion by recounting an incident that happened shortly after the Initiation:

> 'The reactionary police killed two responsible comrades in this area. At that time informers gave information to the police about who the comrades were and their role in the Party's work. Our Party then took revenge on these spies and they were killed. After this, the police went into the village and arrested about 100 people, including me. Other teachers in this room were also rounded up. The police went through the village, looting many things, like knives, cooking pots, sickles, clothes, and musical instruments. All of us refused to give the police any information and so we were brought back to the police post and brutally beaten and tortured, both mentally and physically. They took us outside and when we refused to talk they fired their rifles near our heads and asked us, "What do you wish to tell your family? You are going to die in the next minute."'

Another teacher in the room who was also arrested continues: 'I was taken to the district headquarters and put in a room where they mentally tortured me. They brought the bullets from their rifles to show me and threaten me. Police kept coming into the room and saying, "I will shoot you, starting with your toes and go up and up your body until I reach your chest." They wanted me to make false statements. After 30 days we were brought to the district court and put in jail. Some people were released after paying a high bail. Some were accused of being dacoits [bandits] and killing the spies. Many people were charged in this same case.'

Government repression has changed the whole way these teachers must do their political work. One teacher says, 'Now we revolutionary teachers have to do our political work underground. We are secretly involved in the All Nepal Teachers Organization (ANTO), which is underground. And we are also members of the Nepal National Teachers Organization, which is legal and open. All the teachers in the ANTO are in close contact with the Party. It is hard to work secretly because the police know about us from the past, but we still do work to help the Party. On the one hand, we are active in the teachers' organization. On the other hand, we are Party members and area committee members. We collect donations for the Party from other teachers and also work in other ways to help the Party. We are always on the lookout in the village to be able to give information to the Party on the daily situation. There have been different incidents in this district where the police have killed teachers and students. So we try to work legally until the police know about us, but then we have to go underground and become full-timers.'

* * *

Pravat, my translator, also used to be a teacher in his village. He tells me he taught high school kids during the day and farmed his land before and after class – from dawn to 10:00 am and 3:00 pm to sunset. He had a small plot of land, a two-story house, and several buffaloes, goats, and chickens. He was not rich, but his land and animals provided food and clothes for his family. Soon after the People's War started he had to give this all up.

A short, slender man in his forties, Pravat has been underground now for over two and a half years. One night, as we sit outside under a full moon, he tells me how he went from being a simple peasant farmer and schoolteacher to an underground full-time Party member. The first thing he tells me is how proud he is of his wife. She is also a full-timer in the Party and has been underground since the first day of the Initiation, when she participated in an armed action. She is now a leader in the revolutionary women's organization.

After his wife went underground, Pravat continued to live in their house, taking care of his young son and daughter by himself. But about six months later, the police came and arrested him. They bound his hands and dragged him to a police post quite a distance away. They threatened and questioned him, trying to link him to an incident that had just happened in the area. This was during the monsoon season, and at night they left him tied up standing outside in the pouring rain, wearing only his shorts, a tee-shirt, and rubber thong sandals.

Eventually, Pravat was released, but the police filed several charges against him, including 'public disturbance' and 'treason,' so he was forced to go underground. Later they filed more charges against him, including a false charge for murder, which he knew nothing about. Now it is very dangerous for him to go anywhere near the area where he and his family used to live.

Pravat's house now stands empty. After he left, the police came looking for him and when they found no one at home, they wrecked everything, smashing the cupboards, beds, and other furniture. The two children now live with a relative in another area and see their parents every three or four months for a day, or only a couple of hours. The children have to be careful – they cannot talk about their parents, so they call the relatives they stay with 'mama' and 'papa,' and they call their real parents 'auntie' and 'uncle.'

Pravat says his children know that their parents are fighting in the People's War and his son already talks about how he wants to learn how to use a gun so he can join the revolution. Pravat and his wife work under different conditions and in different areas so they don't see each other often and sometimes only for short visits. But they seem to have a deep bond.

I am struck by the reserved and matter-of-fact way Pravat has of telling the most hair-raising stories. But even when we are joking around there is a serious edge – a side to Pravat that is always tense, alert to the danger in his life, and solid on the commitment he has made to this revolution.

Carrying the Story Forward: Children in the War Zone

Out of Nepal's 23 million people, almost eleven million are under 16 years old.

Nepalese officials, newspapers in Nepal and internationally, and various human rights organizations have claimed that the insurgency in Nepal is responsible for the deaths of many children and accuse the Maoists of 'recruiting child soldiers' and using children as 'human shields.' There have also been widespread reports that thousands of youth have fled their homes in the countryside, allegedly to avoid being 'press-ganged' into the People's Liberation Army.

The official policy of the CPN (Maoist) is that no one under the age of 18 is allowed to join the People's Liberation Army, and minors who have responded to recruitment calls have been told they cannot join the PLA and people's militias. At the same time, the CPN (Maoist) organize youth under 18 to support the People's War in many other ways. An article in *The Worker* (an official publication of the CPN [Maoist]), says: 'While they have been strictly forbidden to join people's armed force, they [the minors] have been organized under *Akhil Bal Sangathan*, a children's organization which takes care of the overall development of children, including their right to express their solidarity to what they consider is good, including the People's War ...'[1]

In February 2003, a report by the Child Workers in Nepal (CWIN) Concerned Center, widely cited in the press in Nepal and internationally, asserted that: 'The death toll for children under 16 has reached 96 as a consequence of the war between the Maoists and the government' and 'nearly 3,000 children have been displaced from their homes and at least 1,500 have been orphaned following the conflict.'[2]

News articles about 'children being harmed by the Maoists' assert statistics like: '168 children have been killed in the Maoist insurrection.' But such reports don't actually say these children have been killed by the Maoists. In fact, the government's own statistics show that it is the police and Royal Nepal Army that have been responsible for killing thousands of people, including many children.

It was widely reported that, by December 2002, more than 7,000 people had died in the conflict between government forces and Maoist guerrillas, and that over 4,000 of these deaths had occurred since November 2001. The vast majority of these deaths were at the hands of the police and RNA soldiers. Government sources say that of the 4,366 people killed between November 2001 and December 2002, 4,050 were Maoists. But as many human rights groups have pointed out, most of these victims were civilians targeted for their real or perceived support for the Maoists. Human Rights Watch reported that in the first few months after the State of Emergency was declared in November 2001, over 1,300 'suspected Maoists,' including 'civilians once associated with Maoists as well as those possessing Maoist literature,' had been killed by government security forces. Between November 2001 and October 2002, 4,366 people had died in the conflict, compared to around 2,700 deaths in the previous five years.[3]

In other words, more people had been killed by the police and RNA in this one year than the total number killed in the first five years of the insurgency.

Those arguing that the 'Maoists are killing children' fail to mention that even those human rights groups that reported that '168 children had been killed in the Maoist insurgency' also reported that government forces had unjustly killed many people, including children, using the pretext of 'skirmishes' or 'encounters' with rebel forces.

A lot of human rights groups, either consciously or not, help spread disinformation and confusion about the situation in Nepal by 'evenhandedly' criticizing the government and the Maoists for 'human rights violations,' even though by their own statistics, the overwhelming majority of those killed have died at the hands of government forces. But these groups do cite and chronicle many cases in which the state has killed children accused of supporting the Maoists. They have also reported on the abuse of children held in jail on suspicion of being rebel soldiers.

One RNA officer admitted that in the heat of battle, government soldiers rarely distinguish between men, women, and children. One army captain told a reporter, 'Anyone with a gun is an enemy.'[4]

Furthermore, any discussion about the plight of children in Nepal needs to look at the semi-feudal and capitalistic system under which millions of children live in dire poverty and brutal servitude. This is the very system the Maoists want to do away with.

For example, in Nepal, there are 32,000 child laborers working in 1,600 stone quarries. Almost half of these children fall ill soon after starting to work and regularly suffer from coughs, backache, fever, visual impairment, and joint and muscle pain. Almost all of them have had accidents and injuries while working to excavate and extract stones and boulders from quarries, loading goods on trucks, or crushing boulders into gravel. One news article recounted the story of a 13-year-old boy who goes to school in the morning and on the way home stops at a quarry site and crushes stones for hours. He earns between 20 and 30 rupees a day (less than 50 cents), which helps his family of five survive.[5]

A nationwide study by Tribhuvan University in Kathmandu reported that more than 27 percent of the children in Nepal – some 2.6 million children – work as child laborers and that 60 percent of the children are between 6 and 14 years old. Almost one million work without pay and many work as bonded laborers, forced to work for an employer for a specific period of time, without any rights.

Deep poverty is also responsible for the suffering of millions of children in Nepal. It is estimated that 50 percent of the children in Nepal are afflicted by malnutrition. And because of the lack of clean water, sanitary conditions, and health care in the countryside, many children die of common, curable diseases.[6]

14
Martyrs of Rolpa

This morning, I decide to sip my morning milk-tea outside. I walk a distance from the house to grab a few minutes of solitude to enjoy the softness of the early morning light and think about how much I like living in the countryside. Every day when I wake up, I feel like I've stepped into an amazing painting. The scenery here is absolutely breathtaking. And while the trekking is difficult, it is hard not to feel energized and exhilarated by the clean air and the towering mountains that encircle us. The natural beauty of the landscape seems inappropriate for the ugly poverty that surrounds us. But the drama of Nepal's rugged and spectacular landscape seems fitting as a haven and cover for a People's Army that is defying all odds.

Today our trek starts just before noon and for hours almost the entire trail is either straight up or rocky downhill. Then just as the sun starts to near the horizon, we come to a very steep mountain where there really isn't any path. We just climb directly up the mountainside – and it seems to go on forever.

We are now at an elevation of about 12,000 feet and it's difficult to get enough oxygen to my lungs and muscles. I get out of breath very quickly and have to stop every 50 yards or so to catch my breath. The people I am with see I'm struggling and one of them offers to literally carry me up the mountain on his back. But I push on. We reach the top of the mountain after nightfall but a full moon has taken the sun's place, so we can see pretty well. Still, the path is difficult because it is all downhill, strewn with loose rocks that seem to have an unpredictable life of their own.

Today, our time on the trail is lasting longer than planned because there aren't any villages in this area. So we have to walk until we find a place where there is at least a water tap. Finally, at 11:00 pm we stop and make camp at an old, abandoned cowshed. I immediately plop down on an old plank of wood. But the squad quickly sets to work. Some leave to fetch water, others gather tree branches and leaves for our bedding, and the rest start clearing out the shed and making a fire.

Here in the western mountains it gets really windy and cold at night. So as soon as the fire gets going, everyone quickly huddles round to get warm. Then we eat our dinner – crackers and instant noodles, eaten dry, straight from the package. Everyone is in a good mood, joking around, and it is past midnight by the time we lie down to sleep. The bed of leaves is actually very comfortable. But 5:15 am comes quickly, and I'm roused from my sleep to get ready to go. We are back on the trail by 5:40 am.

We are now getting very close to where we will cross the border into Rukum. But before leaving Rolpa, we are scheduled to stop at one last village where people have organized a mass meeting to greet us.

Mid-morning we come around a bend in the trail and Pravat points way across to the other side at the mountain facing us. Nestled in the dense green are tiny spots of red. At first I'm puzzled by this decoration, which doesn't look at all natural, and then I realize it is red flags dotting the area, anticipating our arrival.

When we get to the clearing on the other side, people greet us down the hill from where the mass meeting is being prepared. I look up at the 'red dots,' which are now clearly red flags waving lazily in the breeze and I see that villagers are already starting to gather. The sun is high in the sky now and it's getting hot so we rest under some skinny trees, maneuvering to share the meager shade. The squad disappears into the nearby forest to change into their uniforms. I start to doze off a little, lulled by the combination of warm sun and tired muscles. But I wake up when a group of people come walking down the hill, lugging huge pots of food for our mid-day meal.

At 1:00 it is time for the program. We start walking up the hill toward the gathering place, which has been decorated with all kinds of greenery and flowers. A 'doorway' for our grand entrance has been constructed of branches and about 20 guerrillas are lined up in formation forming a corridor for us up the hill. A cultural team is standing at the top, playing music as we march towards the gathering.

I am at the front of our procession and go through the entrance first. About 700 people have gathered and as my head pokes through the doorway, they erupt into loud applause and cheers. A young man steps forward to greet me and puts the traditional Nepali red tikka on my forehead and places a garland of flowers around my neck. Then someone else steps forward and puts a bamboo pole in my hand. The pole is weighty and I can immediately feel its height. I look up and see that it's about 40 feet high with a red hammer-and-sickle flag at the top. I am guided over to the front of the crowd where, as the 'chief guest,' I am given the honor of planting the flag in the ground to start the program. This sets off an even louder round of applause and cheers.

Like every revolutionary gathering here, this one starts with a minute's silence for all those who have died in the revolution. The program lasts for about five hours; there are many statements and speeches from different mass organizations and Party leaders, there are reports from the battlefield, and the cultural squad entertains throughout with their repertoire of songs and dances.[1]

The program ends just as the sun is setting and I am told that several families have come to meet me. Some of them have traveled quite a way and they have all come to tell me the story of how a loved one was killed by the police. The first villager who sits down to talk is 57-year-old Jokhi Budha. Her husband, 65-year-old Singh Budha, her daughter, 22-year-old Kumari Budha, and her son, 29-year-old Danta Budha, have all been killed by the police. She says:

> 'It happened in November 1996. A spy in the village snitched to the police and they came to our home at night. The spies and police arrested my husband and my daughter. My husband was a sympathizer and my daughter was active in the women's organization. Both of them were taken to the police post of the village and tortured for two days. The police then took them to a streambed and shot them with three other people, 70-year-old Bardan Roka, 45-year-old Bal Prasad Roka and 49-year-old Dil Man Roka.
>
> 'Before killing Kumari the police plucked out both her eyes and then put kerosene in her hair and set it on fire. Thirteen people were arrested at this time – five were killed, the rest were released. People told me that when Kumari cried out in pain, saying, "Mommy, mommy," the police said, "You criminal, you Maoist, you terrorist." And then two police picked her up and threw her into the fire alive. Now my son, who was a platoon member, was killed in the recent raid on the police post in Dang in which seven police were killed.'

Jokhi is crying by the time she finishes recounting her terrible loss and I have a hard time holding back the tears that threaten to roll down my cheeks. Through her tears, Jokhi tells me, 'Though my husband, son, and daughter have died, I have many thousands of sons and daughters who will take revenge.'

I have talked with many relatives of people killed by the police. And like them, Jokhi seems to have a larger sense of 'family,' which is why she can feel optimistic even though she has lost her husband

and two children. In traditional feudal society, a woman who has lost her husband would, in most cases, immediately face economic hardship. Older parents who lose their children usually forfeit the security of knowing someone will take care of them in their old age. The Party and the People's Army say relatives of people who die in the revolution are taken care of. Funds are collected among the villagers to give to them. They are given a share of seized land. And children whose parents have been killed are also taken care of by the larger community of supporters of the revolution.

Dil Man Roka was killed in the same incident with Singh Budha and Kumari Budha. His wife, Man Maya Roka, has also come to talk to me and says: 'The people now call our VDC (Mirul) the martyrs' VDC. The police arrested and killed my husband because of his good work. He was killed by the enemy and now the whole Party is here to help me and my family. Now I am farming with the help of the neighbors and taking care of my family. At the time of his death I was four months pregnant with my last son and my other three daughters knew their father. They know the police killed their father. My point of view on the Party is clear. The children's father started to dig the road and the children will finish it. And we will get victory over our enemies.'

Soon after the People's War started, the government struck back with savage cruelty. Paramilitary forces and specially trained commando forces were deployed in large numbers with direct orders from the highest offices of the central government. There were a series of killings in what the guerrillas call 'fake encounters' – incidents when the police kill people and then falsely claim it was an 'armed battle.' In the first three months after the Initiation, more than 30 people were killed by police, most of them in Rukum, Rolpa, and Jajarkot.

I am told that, in addition to shooting people, the police have carried out arrests in the thousands, tortured people in custody, gang-raped women in the villages and in custody, and looted and set fire to people's homes. Because of this, thousands of people have been forced to hide in the forests and caves for months and many have been underground for years. People say that in some places the government has unleashed armed local goons against people and that some villages have faced months of living under curfew.

In the first eleven months of the People's War more than 70 people were killed, while about 40 people considered reactionaries – including local tyrants, police informers, and police – were killed. Of the revolutionaries killed, nearly 40 percent were Party members, more than 60 percent belonged to the oppressed nationalities (most of them Magars from the Western Hills), about 10 percent were women, and almost all of them were of poor or lower-middle peasant class origin.

Government repression really escalated in the third year of the People's War. In one eight-month period, 500 people were killed by the police and many more were jailed, tortured, and raped.

Khala K.C., a 23-year-old woman, was killed during the government's Kilo Sera 2 campaign, which went on from mid-June to August 1998. Her 31-year-old brother, Chitra Bahadur K.C., tells me:

> 'My sister had been working in the revolutionary student organization since 1991. In 1995 she joined the Party and became a full-timer. At the same time she took the responsibility of working in the women's organization. In 1997 she became a squad member. She was five months pregnant when she was killed. She came to the house to visit our sick father and afterwards she took shelter at another house. There were two other comrades at the house – one was her husband, also a squad member.
>
> 'The police came to the family's house and arrested our eldest brother and beat him. There were many police and they surrounded the house. None of the family could inform the people in the shelter about the police. Our eldest brother made a lot of noise when he was beaten and the people in the shelter heard his cries and ran away. My sister fled to hide in the cornfields, but the police found her there and arrested her. They took her to the forest and beat her on the way. Then they killed her. After 23 days her dead body was found and it looked like she might have been raped.'

Sharpe B.K. was 43 years old when he was killed in 1998. He left a wife, Rupsari B.K., four sons and one daughter. Rupsari tells me:

> 'Our family is a poor peasant family and my husband worked in the poor peasant organization. We are of the lower caste – according to Hindu religion, 'untouchable.' A spy in the village told about my husband's activities. Twelve police came to our house and arrested Sharpe. Then another 14 police came and they all took him to the forest. I followed them, crying and begging the police to let him go. But they beat me viciously. I went back home and I heard the sound of gunshots and I thought they must have killed him. I couldn't go to the spot because I was so hurt from the beating.

'After two hours my sons and some Party comrades went to the jungle and found my husband's dead body. They didn't take the body back to the village because of the heavy repression, so they left it there for eleven days, covered with stones and dried leaves. Later the people gave my husband a funeral procession and ceremony to burn his body. The police continued to threaten our family. But now the people's power has grown so the police have been forced to stop harassing our family.'

Nil Bahadur Oli was 21 years old. His younger brother, Purna Bahadur Oli, was only 19 years old. They were both killed in 1998. Their father, Man Bahadur Oli, tells me his two sons remained stalwart in the face of death:

'Nil worked in the YCL [Young Communist League] and peasant organization. Purna too worked in the YCL. Nil was also in the process of becoming a Party member. My cowshed is in Sallyan and my two sons were there. After the action against the Jhimpe Communications Tower, the People's Army took shelter in that cowshed. The tower was guarded by the police – 18 in all. When the guerrillas attacked the tower the police surrendered and one was killed and two were seriously injured. Eight rifles, one revolver, and 780 bullets were captured. The next day police came into the area by helicopter and bus. A large number came to the tower. The police arrested one person from Sallyan who knew about the action. When the police tortured him, he gave them information, including how the squad had taken shelter in the cowshed. Then the police came to the cowshed and arrested my two sons and took them by helicopter to the police post, very far away.

'The police interrogated my sons but they didn't say anything, even though they knew all about the action. They had some Party documents and weapons hidden in the village, but did not tell the police about these. The police tortured them for two days. When they couldn't get any information from them the police took my sons to the forest by helicopter and killed them. For five days the police stayed by the bodies, trying to ambush anyone who might come to get the bodies. Then they returned to the police post. On the sixth day the Party led a funeral procession of about 100 people.'

Man Bahadur Oli tells me all this with deep sadness in his voice, but he seems proud of his sons. As with other families who have lost loved ones in this revolution, what strikes me most about Man

Bahadur Oli is not so much his grief but his unwavering strength and determination. The government hopes that by brutalizing and murdering Party members, guerrillas, and supporters of the People's War they will be able to crush this revolution. But from what I see and hear such vicious repression has only deepened people's hatred for the government and made them feel even more strongly that the only way they can be free is to overthrow the present regime.

15
Families of Martyrs: Turning Grief into Strength

Today we leave our shelter late in the afternoon. The group that has been accompanying us in Rolpa will not be crossing into Rukum with us. We will have to say goodbye to the squad that has been traveling with us, and the people who have been providing political and tactical leadership for our trip are also returning to their villages. There is kind of a ritual people here do when saying goodbye. When it's time for us to go, people initially leave with us, walking part-way up the first mountain to give us a kind of send-off. Then we all stop to say our last farewells.

We cross the border into Rukum and when we arrive at our next shelter it is close to sunset. There are about 100 families in this village, and people have been waiting for our arrival since morning. By the time we've rested a while and had a meal it is after dark. But people gather outside for a program where local villagers and a cultural squad sing and dance, providing a festive ending for my first day in Rukum.

The next day I meet all morning with some local area leaders. But as we have been traveling non-stop for days, we decide to take a break at noon so I can wash my clothes and clean up. Since we crossed into Rukum, a group of women have been traveling with us, including Sunsara. She doesn't speak any English, but through eye contact and gestures, she has been acting in a very motherly way towards me. I can see that she wants to communicate so badly – as I do too. And she repeats a phrase that I have heard from others – 'If we could speak the same language we would be talking for a very long time, all night, sharing our experiences.'

Sunsara helps me clean my clothes and body. One of the women gives me a sari to wear – a brightly printed, very wide piece of material that is wrapped around like a dress. Saris are traditionally worn by Nepali women and they also provide a kind of 'shower curtain' for taking a bath at the village water tap. Today it's overcast and even though the sun is high in the sky the wind has a slightly chilly edge.

The cold water hits my body, sending me into shivers. But it feels good to get clean after days of dusty travel.

After we finish I ask Sunsara if she will sit down and talk with me about her life. I am surprised when she tells me she is 50 years old – I had guessed by looking at her that she was much older. But people's lives here in the countryside are so hard and the impoverished conditions of life – backbreaking work, bad nutrition, and lack of healthcare – take a huge toll on the physical state and appearance of the people. This is particularly true for women who have borne and raised many children.

I am constantly amazed at how strong people here are – even small children can lug very heavy loads up and down mountain paths for hours. But the overall poor health of the people makes them very susceptible to all kinds of disease and sickness. Right now in the Western Region, in Jumla and adjoining districts, there is a major 'flu epidemic that has already struck over 10,000 people and killed more than 400. If you got this kind of 'flu in the US, it might mean a few days off work or school. But here, since people's general health is so poor, it can mean death – especially for older people and young children. The 'flu epidemic has already been going on for a couple of months, but the government has done little to send medicine to the affected areas. I can now see why the average life expectancy in Nepal is only 55 years.

Sunsara, who is active in the local women's organization, starts off by telling me how she became active in the People's War. She says:

> 'I am an illiterate woman who cannot read and write at all. Seven years ago, when my son was seven and my daughter was five months, my husband died. After this I lived in my house for five years with my children. A few years ago some Nepali Congress goons killed two sympathizers in our village and the reactionaries accused sympathizers of the People's War of doing the murder. The police came to the village and arrested over 20 people. They came into the village three or four times a day looking for people, so many people left the village, including me. My children are now with my uncle and there is no one living in my house. I see my children every few months. Sometimes the police come to the house looking for me.
>
> 'Now I'm working in the main committee of the women's organization in this area and I'm also a Party member. I started working in the women's organization two years before the Initiation. I am very happy to be working in the Party because I did not get to study in school

but now I'm being educated in Marxism-Leninism-Maoism. And I feel collective life is happier than living an individual life. When I lived in my home I met many comrades and talked with them. We shared our happiness and sorrow. Now I am committed to sacrifice in every way to liberate our class. Before the Initiation I was very oppressed – on the one hand, by the government, and on the other hand, by the men in the family. All the housework was done by women. Since the Initiation there have been many changes. All household work is now done by men and women. Beside this, men inspire us to go forward, fight to liberate women, and participate in the People's War.

'When I visit my children I tell them, I want to live with you but that is impossible because if I did the police would come here and arrest me. So it is better to do work in the Party than live in this house. It is my duty in this situation because the People's War is growing day after day. We are all involved in the People's War to get victory. And you will also be involved in the war when you grow up.'

* * *

I have been traveling through the districts hit hardest in the last three years by government efforts to crush the People's War – Sindhuli and Kavre in the east, Gorkha in the Middle Region and now Rolpa and Rukum here in the west. In all these places the families of those killed by the police provide stark testimony of the cruel nature of the government's counter-revolutionary campaigns.

On May 1, International Workers' Day, there is a program to greet me in the village where we're staying. Afterwards, several family members whose relatives have been killed by the police line up to tell me their stories. The first to sit down and talk are two women, adorned with nose rings and dressed in brightly colored clothes.[1] They are here to tell me about Kami Buda, who was 27 years old when he was killed in 1955. Moti Kali Pun, his daughter, was only three years old when her father was killed. Kami Buda's sister, Aas Mali, who is now 70 years old, begins the story:

'Kami Buda was in the Indian military. After the Communist Party was established in Nepal in 1949 he left his job in the army and came back to Nepal and joined the Party. In 1952 the Communist Party was banned. When he was here in the west there was a sharp contradiction with liars and cheaters [oppressors in the villages] in Rukum. He was involved in Party activities and the government arrested him in Jumla. At that time

27 people were arrested, including him. He told the authorities he was a leader in the Party, but that the others were not. Then the others were released, but Kami disappeared, he was killed. When he was in jail he was mistreated and he demanded better treatment. This angered the authorities and they took him to the forest and killed him. He could read and write and he wrote articles and poems. He was in jail one month before they killed him. First the government cut off his hands in jail so he could not write, so then he used his feet to write. Then the police cut off his toes so he couldn't use his feet to write. He was the first communist martyr of the Party in Rukum.'

In many cases people have just 'disappeared' – the police take them away and they are never seen again. Fifteen-year-old Naina Shing Chhinal, who comes to tell me about his 32-year-old father, Bhairam Chhinal, says, 'My father was a strong sympathizer involved in Party activities. He was a district advisory committee member of a lower caste organization – the caste that sews clothes. After the Initiation the police called him to the police post many times. He was not charged with any crime, but they asked him about Maoist activities. He didn't tell them anything. In October 1998 he went to the police post and he was disappeared. Formally, family members don't know what happened, but we are sure he was killed.'

In other places I had heard about how spies and snitches give information to the police that leads to the capture and murder of Party members, guerrillas, and sympathizers. And many times, these snitches have been members of the parties the Maoists consider revisionist, phony communist – like the UML (Communist Party of Nepal [Unified Marxist Leninist]). This has also happened in this area. Pawn Kumar Bohra was only 19 years old when he was killed in October 1998. His 30-year-old brother, Dand Bahadur Bohara, tells me:

'Pawn worked in the student organization when he was a student. After he passed the high school examination he went to college and then later left his studies to become involved in Party activities. He became a full-timer YCL [Young Communist League] member and commander of a people's militia. Six comrades were taking shelter waiting to take a military action. He and another comrade went near the police post and the house of a liar and cheater to survey the situation. At the time a UML goon knew about where the comrades were taking shelter and went to the police post and brought the police back to the area. The police saw the two comrades walking on the way and attacked them, shot at them. One of the comrades was able to run away but my brother was killed. The other comrades in the shelter heard the sound of gunfire and were able to run away and escape.'

The murder of Obi Ram B.K. is another case where people think a spy might have snitched to the police. Obi was killed in 1998 at the age of 23 and his father and mother, Mohan Lal B.K. and Pabita B.K., sit down next and tell me about the death of their son.[2] Mohan says, 'Our son was underground since the Initiation. He worked in the student organization and was also a squad member. He left student work in 1997 and was promoted to squad commander. He went to the district headquarters for a military action with three squad members under his command. They took shelter in a hotel and the police found out about their presence, perhaps from a spy. The police came to the hotel and arrested them. They were beaten and tortured and the police tried to get them to surrender and give information about the Party. But they refused. The police tried for 24 hours to get them to talk and then killed them.'

Some of the most heartrending stories I hear are about young women killed by the police. A lot of these women joined village militias or the People's Army when they were only teenagers. And like most women in Nepal, many of them were not allowed to go to school when they were growing up.

Binita Buda was only 16 years old when she was killed in 1998. Ghiumali Buda, her 63-year-old grandmother, tells me that Binita was the treasurer of the area committee of the revolutionary student organization, a member of the area committee of the YCL, a Party member and squad member. A member of the Party who was with Binita when she died recounts what happened: 'Three of us, Binita Buda, a peasant, and me, went to the village to survey the situation in preparation for a military action. The police knew about our shelter and encircled the village. But we didn't know about these police activities and we went to the shelter. Before reaching the shelter we met the police on the way. There was a face-off, but there were so many police we had to run away. The peasant and I reached the shelter, but Binita did not. Then after some time we heard she was arrested and taken into custody. The police tortured her and raped her the whole night and then killed her. She was the second woman martyr of Rukum district.'

The last relative I talk to today is Shankar Lal Gharti. He is the 24-year-old son of Bhadra Bahadur Gharti, who was killed in 1998 when he was 49 years old. Shankar says:

'My father was a staunch sympathizer of the People's War. Other comrades and I slept in the forest all night, and early in the morning came to my family's house. We didn't know the police were around. When we entered the house my mother asked us to wait to drink tea. When the tea was ready the police came to the house and encircled it. The police fired into the house and another comrade, Chain Buda, was wounded in his left arm. We realized the police wanted to kill us and we shook the hands of all our family members. We said, if the police kill us, they will not kill our ideology and political line. Inside the house we wrote the slogans, "Long live the People's War," "Long live the CPN (Maoist)," and "Long live MLM."

'The police told us to come out of the house and said if we didn't they would burn the house down and kill everyone. So we decided to go outside. First, we sent my mother and when she went out the police arrested her. Then my wife left and they arrested her also. Then my father came out and they arrested him.

'Only two of us were left and we came out last and they arrested us and tied us up. They tortured us, beating us with sticks and the butts of their rifles, and tried to get us to tell about the Party and our activities. We didn't say anything and the police called the district to ask what to do. The order was given to kill us and they stopped asking us questions.

'The police took Chain Buda away and tried to get information from him. He tried to escape and they shot him ten or twelve times and killed him. Then the police took my father away and killed him also. When Chain Buda tried to escape, I also tried to run but the police grabbed me and tied my hands behind my back. I was able to escape after that and run away. When I ran away, after about 60 meters, there were more police who caught me and the others fired at me from behind. I was shot in the back and fell to the ground. The police thought I was dead and went away. Then some peasants carried me to a safe shelter and I was taken to Kathmandu for treatment.'

* * *

By the time I finish talking to all these relatives it is starting to get dark and it's time for our evening meal. I thank all the family members for coming to talk with me and I tell them their stories will be heard by people in the United States and beyond.

I ask if they will let me take their photos and one-by-one they come outside to pose. Each of them stands tall and looks straight into the camera, as if they are trying to send a message through their eyes.

These relatives have given me yet another look at why the government is having such a difficult time trying to crush this People's War. Bhiusan Chhinal, the father of 28-year-old martyr Kal Sing Chhinal, captured the attitude of many people here when he told me, 'My son lost his life in a war for the people and nation. But there are thousands of other sons who will now fulfill my son's aim.'

Carrying the Story Forward: The Rising Death Toll

In September 2002, the commander of Nepal's Armed Services told Amnesty International that the army's mission is to 'disarm and defeat' the Maoists and that the definition of what constitutes a 'Maoist' includes civilians who give shelter, food, or money to the Maoists. A senior superintendent of police also admitted to Amnesty International that security forces deliberately kill 'Maoists' and explained this by saying that the rough terrain and lack of detention facilities made it difficult to take wounded Maoists to hospitals or captured Maoists to prison.[1] In fact, while there are daily reports of 'Maoists' killed in 'encounters' with security forces, there are few reports of Maoists injured or arrested – which suggests that security forces do have a policy of deliberately killing Maoist suspects instead of arresting them. During 2002, Amnesty International detailed the killing of more than 200 people in approximately 100 incidents to the UN special rapporteur on extra-judicial, summary, or arbitrary executions.

One victim of such practices was 19-year-old Sukuram Chaudhary, a watch repairer, farmer, and student, from Pathariya, Kailali district. He was killed by the police on September 10, 2002 along with three members of the CPN (Maoist). The *Kathmandu Post* reported:

> The night before they were killed about six or seven Maoists had come to the village and had demanded food and lodging. The next morning, around 8 am, there was a search operation in the village. The Maoists ran away, leaving some of their weapons behind. Sukuram Chaudhary and his family were eating at home. The security forces started to question him. However, as he belonged to the Tharu community, he did not understand Nepali. They then started beating everyone with lathis [long sticks], kicking them with their boots and slapping them in the face. Then they locked the women and children in the house. They reportedly took Sukuram Chaudhary to a path near the canal nearby and beat him severely. They put sand in his mouth to stop him shouting. After that, they returned to the house and took his wife. She saw her husband blindfolded with his hands tied behind his back lying on the ground. Then she was beaten in front of him and taken back to the

> house where she was allegedly raped by one of the APF [Armed Police Force] personnel. Later that morning, around 11 am, villagers reportedly heard about 20 gunshots coming from near the bridge. It is thought that Sukuram Chaudhary and three Maoists who had been arrested by the APF were summarily killed, and their bodies disposed of nearby. The next evening, there was a news item on Radio Nepal at 7 pm stating that four 'Maoists' were killed in an 'encounter' at Pathariya.

This same article describes how 35 laborers working on construction at the Suntharali airport in the Kalikot district were killed by RNA soldiers. This incident took place after the February 17, 2002 attack by Maoist guerrillas at Mangalsen, Achham district, in which 56 RNA soldiers were killed. According to the *Kathmandu Post*:

> On 20 February 2002, an army helicopter had tried to land at the Suntharali airport strip but had been shot at by a group of Maoists. After that, the foreman, Kumar Thapa, from Gorkha district and his helper, Prem Lama, stopped the construction work and instructed the labourers to stay in their homes or huts. They collected the identity cards of all labourers. When the army arrived on foot around 9 am on 24 February 2002, Prem Lama tried to show the identity cards but one of the soldiers fired at him. Then the other soldiers dragged the labourers from their temporary homes out and shot all of them.

Women and children are among those who have been unjustly killed by RNA soldiers and police. For example, on July 22, 2002, a joint patrol of about 400 army and police personnel went to Jagatia village, in the Bardiya district. They entered the home of Ram Kishan Tharu and questioned him about providing food to the Maoists. When he denied it, he was hit on the legs. The security personnel then grabbed his twelve-year-old daughter, Rupa Tharu, blindfolded her, took her away, and shot her dead. The police then reportedly took her body away and buried it in the presence of two local politicians. On July 23, 2002, a press release by the Ministry of Defence reported the incident saying a 'Maoist was killed while trying to run away during a cordon and search operation.'[2]

* * *

In January 2002, *Himal* magazine, commenting on the most recent report on Nepal by Amnesty International, said: 'The fact that 73 percent of nationwide insurgency-related deaths have been caused by the state has led many to question the tactics being used; and the impunity with which security forces are operating is also highlighted in AI's report as a "longstanding problem".'

The *Kathmandu Post* reported on January 20, 2004 that 7,792 people had been killed in the Maoist insurgency and, according to a spokesman for the Home Ministry, in just the past four and a half months over 1,300 people, including 225 civilians and 900 rebels, had been killed.

16
Women Warriors

When we cross into Rukum I have to say goodbye to Rachana,[1] who had been constantly with me as my aide while we traveled through Rolpa. In the dark of night her outstretched hands had pulled me up the steepest parts of the mountains and steadied me when I teetered, crossing dangerous ravines. Like most guerrillas in the People's Army, Rachana is young and from a peasant background. For long days and nights she treks up and down the mountains, as quick and sure-footed as the men in the squad, carrying heavy loads, along with her rifle.

One day I asked Rachana if she would tell me about her life. At first, she hesitated, surprised that I wanted to interview her. Then she said, 'OK, but let's do it later, after dinner. First, I want to think about what I want to say.' As it turned out, it was a couple of days later – when we reached the border where we had to say farewell – that we found an opportunity to sit down and talk. Rachana had seen me interview Party leaders and military commanders and now looked eager and excited to be the one to have her words written down in my notebook. When I asked her to tell me something about her family and what it was like to grow up as a woman in her village, she said:

> 'There are eleven members in my family – my mother, father, three brothers, two sisters, my brother's wife and three of my cousins. I am the oldest daughter, 18 years old, and I come from a peasant family in Rolpa. My mother and father allowed my three brothers to go to school but would not let me. They told me it is worthless for a daughter to study because she will just get married and move into another household. At the time, this made me feel very sad. When a six-month adult education class opened in the village, I went to learn to read and write. But when I did this, my father would always tell me not to go and he would order me to go to the forest instead, to cut grass and collect firewood.'

Rachana went on to describe how she came to join the People's Army:

'Before the Initiation of the People's War I did not know anything about politics or parties. But after the Initiation one of my relatives suggested that I take part in the local cultural group and asked me to go to their rehearsal. I didn't tell my mother or father about this. I only told my older brother who said, "Go ahead, if you want to die ... Can you carry a gun on your shoulder?" I replied, "You didn't give me a chance to study and now I am eager to solve the problems of the people and the nation. I want to fight for liberation. If you won't allow me to go I will rebel."

'One of the local Party comrades came to talk to my family and he came over several times to discuss revolutionary politics and the People's War. One year ago, after many discussions, my father and mother happily allowed me to join the Party. I started working in the women's organization and was in the women's militia. Then, eight months ago, I was promoted to be in this squad. I am optimistic about the People's War.

'Now all the members of my family are clear on the politics of the People's War. All of them are in mass organizations, and my 15-year-old sister is going to school. She has passed class six and is teaching other people to read. When I was taking the adult education class I never had time to study. But in the People's Army I have time to study reading and writing and the other comrades help me. I can read newspapers and write letters now.

'I was eager to work in the Party before. But then after joining the squad I was involved in an encounter and became even more committed. There were 14 of us going from one place to another and the police ambushed us. One of our comrades was killed and now I have a strong commitment to take revenge. I will fight against the enemy as long as there is a drop of blood left in my body. I am very happy now and we will certainly achieve our goal.'

Rachana's description of how she was denied an education is typical of the way women are treated in Nepal. In the countryside there is a saying: 'To get a girl is like watering a neighbor's tree. You have the trouble and expense of nurturing the plant but the fruits are taken by somebody else.'

Under feudalism a daughter is 'useful' and 'valuable' in her childhood years when she can do chores and serve the household. But according to such feudal thinking, it is not worthwhile to 'invest' in a girl by giving her an education because she will just end up marrying and going to live in, and serve, another household. I did meet a number of women who had been allowed to go to school, at least up until high school. But when I visited colleges in the cities, almost all the students were men.

One afternoon, I watched Rachana studying, practicing her reading and writing, her eyes glued to a dog-eared page in deep concentration. I thought about how this scene is being created in other guerrilla zones in Nepal. Young peasant women – illiterate, facing nothing but a back-breaking future – leaving their villages, taking up arms, learning to read and write, and studying politics. I met many other women like Rachana – women who grew up angry about the way feudal society oppresses women and who had jumped at the chance to join the People's Army. This is clearly another element fueling this revolution.

Everywhere I go, it seems the women are particularly enthusiastic about this revolution. I see it in the eyes of the old women who have suffered many years under the thumb of feudal relations – and who now want to fight for a new kind of society. I hear it in the words of young women who never went to school, who tell how excited they were the first time they carried out an armed action. I feel it in the determination and spirit of the women who have lost husbands, sons, and daughters, but continue to shelter and aid the guerrillas at the risk of their own lives.

These women really believe that the fight against women's oppression is woven into the fabric of this People's War. So when the armed struggle started in 1996, it was like opening a prison gate – thousands of women rushed forward to claim an equal place in the war. Some had to defy fathers and brothers. Some had to leave backward-thinking husbands. Others ran away from arranged marriages where parents had decided their fate. They all had to rebel against feudal traditions that treat women as inferior and make women feel like their ideas don't matter.

In Kathmandu I interviewed Rekha Sharma,[2] the president of the All Nepal Women's Association (Revolutionary) who described the situation for women in Nepal:

> 'In the rural areas women are oppressed by the family, mother-in-law, husband – and some women are killed because of dowries. This problem exists all over the country, in the city and countryside. The thinking in society is that women are brought into the home to serve the husband and to have children, that this is all they're good for. To solve these kinds of problems we try to educate women, to show that it's not because of their mother-in-law, husband, etc., but that it is the social structure that

is protected by the state and that we need complete change, revolution. We educate women to this fact.'

Rekha also told me about the widespread trafficking of women. Every year, 5,000–7,000 females between the age of ten and 18 are 'exported' to India and forced to work as prostitutes. These young women are literally sold for the price of cows and goats by their own fathers, brothers and uncles. Sometimes they are tricked into going with the promise of a 'good job.' Estimates of the number of girls and women working in India as prostitutes range from 40,000 to 200,000. Nepalese women are also kidnapped and taken to countries in the Gulf area to work in brothels. Every year hundreds of women return to Nepal after being forced into prostitution in another country, and many of them come back HIV-positive.

Early marriage, early pregnancy, and multiple pregnancies take a toll on women's mental and physical health. And there is a lot of pressure for a woman to produce a son to inherit the family's property, even if this endangers her health. In fact, Nepal has one of the highest maternal mortality rates in the world – 539 per 100,000.[3] Women who do not produce sons are frequently abandoned, socially ostracized, and many times their husbands take a second wife.

Women in Nepal also face extremely oppressive anti-abortion laws. It is illegal to have an abortion here – abortion is classified as homicide and punishable by law, even if the pregnancy is a threat to the woman's health or life or is the result of rape or incest.[4] Many women are in prison, serving sentences of up to 20 years, for having an abortion. These strict anti-abortion laws have also given rise to illegal, unsafe, and induced abortions which, according to news reports in Nepal, account for more than half of the country's maternal deaths. I read about one doctor who toured hospitals throughout Nepal and discovered that huge numbers of women were in hospital because of complications related to an illegal abortion. In the Maternal Hospital in Kathmandu, 61 percent of gyne-obstetric cases he observed were abortion-related.

The most simple and routine parts of the day here in the countryside are being affected by the People's War in terms of the division of labor between men and women. In feudal society, women are saddled with prescribed roles that keep them in a subordinate position – taking care of children, cooking, washing clothes, etc. But in the guerrilla zones, this seems to be changing. For example, it was very common, as we traveled, for the men in the squad to do a lot of the day-to-day cooking.

Sometimes we get to a village and right away, men from the squad start gathering firewood and preparing the meal. When we sit down to eat they serve the food and then wash the dishes after we finish. Sometimes women in the squad and women in the village will be sitting around doing something else while all this is going on. As I observe all this, I think about how unusual and *radical* this scene is in a semi-feudal country like Nepal. One woman told me:

> 'There have been many changes in people's thinking since the Initiation. Fathers and brothers are now involved in things like cooking, getting water, washing dishes. There is also a change in the women's thinking. Before, women were not permitted to do things like make the roof of the house or plow the fields. But now where the People's War is going on, it is easy for women to do this. Before, women didn't make baskets and mats according to tradition and women used to think they weren't good enough to do this work. But when we dared to do it we found it was easy. So if we dare, we can do anything – there's no distinction between men and women. There are two changes [with regard to the roles of women] – ideologically and in practice.
>
> 'There are two things that have led to women doing "men's work." First there is necessity – some of the men have to go underground and so then the women have to plow the fields, make roofs and do other "men's work." For example, in my father's village all the men had to leave because of the police. The police raided and looted things like food, grain, and ghee [cooking butter]. So the women started to plow the land and make house roofs. The second thing is women becoming ideologically convinced to do such work.
>
> 'Before the Initiation few women dared to do "men's work," but after the Initiation there is no type of work women don't dare do. Also, men can do any type of "women's work" and they don't hesitate to do it. Before men didn't think this way, but with the Initiation their ideology changed.'

I also notice other ways this revolution seems to be changing the way men and women relate to each other. For example, living and traveling with the guerrillas, I am somewhat surprised by the way the women and men are completely relaxed around each other. There is no 'sexual tension' and I don't feel nervous or unsafe like women in the US often feel when they are alone with men they don't know very well. Sometimes we find ourselves in a situation, like in a cowshed or on the floor of a peasant's home, where we all have to sleep together like sardines in a can. But I never feel uncomfortable in these situations.

In all the guerrilla zones I visit, it seems that the women still have primary responsibility for taking care of the children. But this is also starting to change slowly. I have met many women Party members with small children, and other people are always taking turns caring for the children. In the 'revolutionary community,' everyone is considered an 'auntie' and 'uncle' to the children of martyrs or kids whose parents are absent because of their revolutionary responsibilities.

There is not yet organized collective childcare. But several party leaders told me they are trying to figure out how to set this up so women can play a bigger role in the revolution. When base areas are formed where the guerrillas will have (relatively) more stable control, it will be easier for them to organize and maintain things like collective childcare. But at this point, most of the women involved in the revolution full-time have to take their infants with them wherever they go. They do this while nursing and then, when the baby gets older, they find a relative to take care of the child.

For many women, the People's War offers an immediate escape from an oppressive situation where they can't go to school, may be forced into an arranged marriage, and are expected to spend the rest of their lives devoted to husbands, in-laws, and children. Many women find a new life in the revolution – women who have been abandoned by their husbands, who have been socially shunned because they were raped, or women whose family could not afford to pay a dowry for them.

The women who have joined the revolution feel strongly that the present government will not and cannot do anything about the inequality women face. And they are inspired by a revolution that puts forward the vision of a society in which women participate equally in all aspects of life.

One woman organizer in Rukum tells me:

> 'There are various reasons I became a revolutionary. First, there is inequality between sons and daughters – like in terms of property, daughters have no rights. Women get neglected compared to men, by parents, husbands, and other family members. Nepalese women are suppressed by the feudalistic system and some women go to India to

become prostitutes. This women's oppression is the main reason why I was inspired to become a revolutionary.'

I also heard stories of the obstacles women encountered when they wanted to join the revolution. One leader in the revolutionary women's organization tells me:

> 'My father's brother was the head of our household and at first he wouldn't give me permission to join the women's organization. I rebelled against this and for six months I lived somewhere else. When I went back to the house the family members would not accept me because of what I was doing. The women's organization, including me, went to my uncle and tried to convince him, talking to him about women's rights, and we did this many times. He did not speak out against the women's organization but he still didn't want me to participate. He wanted me to stay home and do all the work in the house.'

Another woman in Rolpa says:

> 'At a young age my family arranged a marriage for me. I went to live in my husband's family house when I was 15 and lived there for eight years. My husband was one or two years younger than me but after two years he was sent to India to work and didn't come back. I went with my brother to India to try and find him and we brought him back. But then, within a month, he went back to India and never returned. I am 27 now. Two years ago I left my first husband's house and married my second husband – a love marriage, not arranged. When we got married he was a squad member and he is now a platoon member. In the war period it is easier not to have children, especially for full-timers. And even after the revolution it will be better to have fewer children than more.'

In Rukum I talk with the chairman of the District Committee of the women's organization. She is a Party District Committee member and her husband has been in jail for the last two years. She tells me:

> 'After the Initiation more and more women became involved in armed struggle, in volunteer groups, militias, and squads. In this area, there are eight women's militias with five to seven members and there are also women in militias made up of both men and women. Women are involved at different levels of the Party, up to district Party committee

members. In the whole district there are about 500 local committees of the women's organization, eight area committees, and one district committee. They are all active and they participate in the revolutionary united front.'

The People's Army has a policy that each guerrilla squad (which consists of nine to eleven members) must recruit at least two women. Women guerrillas work as combatants, do propaganda work, and farm the land. Women not directly involved in fighting work as organizers, propagandists, cultural activists, and nurses. They do logistics, spy on the enemy, provide shelter for Party cadres and guerrillas, and visit families of martyrs and those in jail.

When I think about how these women grew up, suppressed by all kinds of feudal traditions, I realize how hard it must have been for them to join this revolution – often in defiance of their families. The sight of young peasant women carrying rifles, khukuries, and grenades is one of the strongest images I carry with me as I travel through these guerrilla zones.

17
New Women, New People's Power

We have been in this village for just one day, but we are scheduled to leave this afternoon to continue our journey through Rukum. The kids have been playing volleyball in the schoolyard but late in the day, the area is cleared and prepared for a mass farewell meeting. Everyone in the village comes out for the ceremony and the atmosphere is very festive. The local militia, made up of about a dozen young men, line up in the yard and give me and my translator a salute as we walk to the front where a long wooden table has been set up.

There are brief speeches and the local cultural group performs. Then leaders of the local mass organizations step forward to present me with gifts – cloth woven in the local villages, two handmade bags, one with the name 'Rukum' embroidered on it, and a traditional knotted rope used to carry heavy loads strapped to your forehead.

The sun is low in the sky when the meeting ends and our entourage gets ready to leave. All the villagers line up along the way to say goodbye and the local militia leads the way, giving us a serious escort.[1] As we head out on the path, everyone waves and shouts *'lal salaam.'*

It is dusk as we leave and the sun disappears behind the mountains soon after we set off, so we have to travel in the dark. After several hours we aren't anywhere near a village where we can find shelter, and the people I am with tell me it will take about four more hours to get to another house. We decide to sleep out in the open tonight and luckily, we come across some shepherds who offer us a place by their fire.[2] They are traveling with a big flock, and we spend the night sleeping amidst barking dogs and baa-ing sheep, milling about only a few feet from our heads.

The next day we arrive in another village in Rukum where some relatives of people killed by the police have gathered to talk with me. As I hear their stories it becomes clear that women here are increasingly playing an equal role in making sacrifices – fighting and dying in this revolution.

In the first three and a half years of the People's War more than 800 people were killed – nearly 100 of them women. Many women have lost their husbands and sons. Many women have been murdered, raped, and brutalized by the police. I have been told numerous stories of women who have remained defiant, even in the face of torture.

One woman organizer tells me:

'After the Initiation the reactionaries put a lot of effort into trying to stop women from participating in the People's War. I'll give you one example. During a local election the police came to ask women to participate in the elections. But the women refused. So the police rounded up more than 14 women and raped them all in one place. There was one twelve-year-old girl that they raped who was so badly injured she could not even walk for a week.

'In some cases women have been "disappeared" and sometimes, the police will try to force women they have arrested to marry them. There is a Nepalese tradition that when a woman gets married, red powder [tikka] is put on her forehead; and the police will do this to the women they arrest in order to humiliate them. They also put arrested women in police uniforms and order them to act like police. But the women have refused to do this and they don't tell the police any information or secrets.

'There have been a number of incidents where women have been tortured in the west. In one area, the police tortured the secretary of the women's organization. She was pregnant at the time and the police raped her. She broke through the police encirclement and tried to escape, but after being beaten she was too weak to run away, and the police shot and killed her.'

Rekha Sharma had also told me about a number of instances in which the police took especially vicious measures against revolutionary women:

'The police go to the homes of people they suspect of fighting in the People's War and they ask the women, "Where is your husband?" or "Where is your son?" In the Pyuthan district, for example, a husband who was a teacher was not home, and his wife, Radha Ghimire, was pregnant. The police came and asked her about her husband, then arrested her and took her to the forest. No one knew what happened to her, and then two to three weeks later she was found beheaded. Another example is Devi Khadka, who was the chairperson of the All Nepal Women's Organization (Revolutionary) in the Dolakha district. She was arrested because she had a pad of stationery with the letterhead of our organization. She was gang-raped in custody. Another woman went to visit her husband in jail and was beaten and later died.'

In Nepal, women's oppression is rooted in the feudal and semi-feudal system of production. The peasants rely on the land in order to survive. But women have not been able to own or inherit land on equal terms with men.[3] Also, in some regions where there are big landlords, peasants face debt slavery. In this situation, where peasants have to work to pay off a debt, some women are forced to provide sex and labor to a landlord.

The revolutionaries here see an agrarian revolution as central to the New Democratic Revolution they are trying to carry out, and part of this is directly addressing the question of women owning land. Land reform is carried out under the slogan, 'Land to the tiller,' applying the principle of 'women's equal right to property.' In the areas where new people's power is being practiced, many widows and unmarried women have had land restored by the people's courts – land that had been stolen by landlords and others who oppress the peasants.

Throughout the guerrilla zones, women are participating in the '3-in-1 committees' that have been organized to oversee and run various aspects of village life. As Party members, as members of the People's Army, and as participants from the United Front mass organizations women play a role in exercising new people's power.

Almost every single peasant I talk with says they could not grow enough to feed their family for the whole year. So for several months each year, the men leave to look for work in the cities or in India, which means the women and children are left to work the land and run the household. Now in areas where the People's War is strong, collective farming is helping households where husbands are either away fighting in the revolution, have died, or have gone to work in the cities.

One woman organizer explains: 'Because there is so much repression, the situation is such that most of the men cannot stay in the villages openly. But the masses of women are still living in the villages. So it is the women who are carrying out much of the new people's power through the 3-in-1 form. The women are dealing with many local problems, like quarrels and disputes among the people. In the 3-in-1 forms, the women make up about 30 to 50 percent of the members.'

When I talk with a Central Committee member of All Nepal Women's Organization (Revolutionary) in Rolpa, I ask her how they are dealing with cases of women being mistreated by their husbands. One example she gives is a situation where a man in one village had cheated on his wife. She says:

> 'The women's organization led the masses of women to punish him by putting a shoe garland [his shoes tied together made into a 'necklace'] on him, painting his face, and parading him around. There was another case of a poor peasant woman who went to work in Kathmandu carrying bricks, and a man from this district raped her. Our women's organization found out about this and went to his home. They arrested him and punished him – they cut off the hair on half his head and made him hold his ears and repeatedly stand up and sit down in front of everyone. We exposed his crime to all the masses. The people support this kind of work we are carrying out in the local areas.'

I also learn that polygamy is a feudal tradition practiced in Nepal. The Party opposes polygamy and arranged marriages and, where the People's War is strong, there is a lot of struggle to put an end to these practices. In some areas the 3-in-1 committees have implemented new policies to ensure the rights of women oppressed by such practices. A woman in Rukum tells me:

> 'In our district there have been many changes in three years in the attitudes and practices regarding women. For example we are involved in the judiciary department [one of the people's power committees] and we take part in solving many problems – like cases of second marriages, where we guarantee the property rights of the first wife. Sometimes the second wife is compelled to leave the marriage because of injustice to the first wife.
>
> 'People who want to get divorced appeal to the people's court and we decide the case. Most are women who want to get divorced because of beatings or attempted murder or other abuses. If a woman demands property from her husband when she gets divorced, the court will help her. The court investigates who is right and wrong in each case.
>
> 'Some first wives are neglected after a second marriage. The woman who gets divorced after a second marriage is now free from her husband. According to tradition, if a woman remarries, the first husband gets some money from the second husband [like a dowry]. But we have now eliminated this practice. In the case of a woman who leaves her husband with no second marriage, there are people who still follow this tradition

> [of the husband getting some money], but the goal is to eliminate this custom completely.'[4]

People tell me that the revolution is challenging many other feudal traditions that oppress women, like the strong preference for sons, the treatment of women as 'untouchable' during menstruation, and the tradition that widows are supposed to mourn for the rest of their lives. Many of the women I met whose husbands had been killed in the war were defiantly wearing bright clothes and jewelry, which a widow is not supposed to do according to feudal traditions. Some of the traditional celebrations are also being transformed by the revolution. For example, the Hindu celebration known as 'Teej' is a day-long fast imposed on women. On this day married women are supposed to pray for the longevity of their husbands and unmarried women are supposed to fast and pray for an eligible husband. Now, this day is being transformed into a day to build support for the People's War.

Throughout Nepal, all the different political forces, including the conservative and reactionary parties in the government, have organized groups of women. But the Maoist women's organizations distinguish themselves by linking the fight against women's oppression with their goal of seizing power through armed struggle, establishing a new democratic state, and moving on to a socialist society. This is very different from the view of other women's organizations in Nepal who work with and within the government to fight for various reforms.

Revolutionary women's organizations in the cities popularize and build support for the People's War. But the Maoist women's groups in the countryside are directly involved in the war – encouraging women to join the People's Army and the militias, building new forms of people's power, and transforming social relations among the people. A woman organizer from the western countryside told me:

> 'In the early days, we started by raising the consciousness of women, like talking about the struggle for equal rights. And we opposed the conservative traditions that oppress women, like the practice where if a woman is married to a man in the countryside, the man's family has to pay a dowry. [In the city, it is the woman's family that has to pay a dowry.] The dowry may be something like some bottles of alcohol, some roti [bread], or goats. So we say that this amounts to the selling of women. So we oppose this kind of practice. Also, in some areas,

we have been able to stop men drinking alcohol and as a result, a lot of wife-beating has stopped. So in a practical way, many women are attracted to the revolutionary women's organization because of these kinds of things. We have also opposed child marriages and polygamy and we've organized mass actions against polygamy.

'We have built a number of *chautari* [memorials] to martyrs and paths in memory of martyrs. We also help build toilets and promote hygiene in the villages. And we support each other in working on the land – organizing to work collectively. We have literacy programs, and we tell women to have fewer children because of the poor economic conditions. Before the political influence of the revolution, men always had the right to decide when to have children. But nowadays [in the guerrilla zones] most of the women and men decide this together. And we say that women should have the right to decide if and when they want to get married and have children.'

The Maoist women's organizations also set up classes for women to discuss the politics and ideology of the People's War. And the women tell me that in many areas in Rolpa and Rukum, there are hardly any women who are not in some revolutionary organization. Members of the revolutionary women's organization are as young as 14 and as old as 70. But the majority are 15 to 30 years old.

I asked one woman, a Central Committee member of the All Nepal Women's Organization (Revolutionary) in Rolpa, to describe the work of her organization. She said:

'Our organization started adult education for both men and women. We have built many monuments to martyrs and we have also built paths for moving cattle. By tradition women go to their parents' home and take tikka on different festivals. But we break this feudal tradition and don't go to take tikka on these occasions. We are establishing a new culture. Instead of practicing these traditional festivals, we celebrate new festivals like February 13, the anniversary of the Initiation of the People's War. For this revolutionary festival we go to meet with the families of martyrs, bring them presents, and take tikka from them.

'If a woman comes to our organization and asks us to punish an oppressor, we help. For example, there was one man who drank and beat his wife so she didn't want to stay with him. She came to our organization with her problem. Her husband didn't want to leave her, but she wanted to leave him. The women's organization decided that she should leave.

> 'We also carry food, grain, and communications for the Party and we do different types of logistic work. We make bullets for muzzle rifles and gunpowder and we set up shops to sell goods to the Party at cost. Local and area committee members make gloves for the squads and platoons. We write the slogans of the Party on the walls. We inspire women to join the militia. We produce food for the Party and also give chickens and goats to the Party.'

Some of the women I talked to said that the men encourage the women to get involved – to develop politically and become full-timers. There are many couples in which both the man and woman are committed fighters in the People's War and when they get married they have a 'communist ceremony' instead of a traditional wedding. One woman organizer told me, 'Before the Initiation [in the revolutionary movement] there was a lot of support for women in theory, but not always in practice. But this really changed after the Initiation and now women are playing a big role in the People's War.'

Carrying the Story Forward: The Fight for Women Leaders

From talking to many women involved in the revolution in Nepal I could see that the revolution was clearly challenging and changing people's feudal and patriarchal ideas about women's roles. But these transformations were clearly only preliminary steps in a long battle to liberate women in Nepal.

When I interviewed Prachanda, I asked him to talk about the problem of developing women leaders in a country where the oppression of women is so deeply built into the economic and social relations. He told me:

> 'Before the Initiation, the woman question was not so seriously debated in our party. That was our weakness. And in our society, male domination, feudal relations have prevailed for a long time. In general terms we agreed, yeah, the woman question is important. As communists we know these things. But in a concrete sense, in a serious sense, I will say that before the Initiation we were not so serious on the woman question. And because we were not serious, therefore, many woman comrades were not at the forefront of the movement. There were some women sympathizers and some organizers, but there was not much effort to develop the women comrades. Then right after Initiation the question came up – it boldly came up. And especially in my experience, I was very thrilled when, during the first year after Initiation, I saw the sacrifice women were making in the main region, in the struggling zones – their militancy, their heroism, and their devotion. When I saw women masses come into the field, then we started to debate seriously the woman question ...'

Prachanda went on to talk about the different problems they confronted in getting women involved and developing their leadership. They were beginning to discuss organizing collective childcare. They encouraged young couples involved in the struggle to put off having children for several years so the women would not end up tied to the home. And they were also trying to deal with illiteracy among women and the lack of birth control.

I thought back on this conversation when I came across a January 2003 article titled 'The Question of Women's Leadership in the People's War in Nepal' by Parvati, a member of the Central Committee of the CPN (Maoist) and the head of their Women's Department. In this article, Parvati talks about the problems the Party is having in developing women's leadership. She says women have joined the PLA in extraordinary numbers and these women have shown much sacrifice and devotion, but only a few have been able to develop as leaders in the military struggle and women themselves are raising questions about the quality of their participation.

I found this discussion fascinating because it was not a romanticized and unrealistic view of the role of women in the People's War and it candidly talked about persistent problems within the Party itself on this question.

According to Parvati, when women get married and have children their participation usually decreases or stops and so, she says, the institution of marriage has 'robbed us of promising women leaders.' While men continue to participate in the PLA, there are hardly any women who stay in the guerrilla ranks after they reach 25 or so.

Parvati says many things work against women getting involved and staying involved in the revolutionary struggle, especially in the PLA, which requires tremendous sacrifice. In the areas controlled by the Maoists there is a struggle against institutions and ideas that prevent women from equal participation in society. Entrenched feudal tradition and ideology – like the view that women should not inherit or own land or that women should be restricted to particular tasks and not allowed to do other jobs – still exert a very powerful force, including among the revolutionaries themselves. Parvati says that there is sometimes covert or even overt pressure on women cadres to get married; unmarried women are treated with suspicion by men as well as women. As a result, some women marry against their wishes or before they are really ready to get married. And there is still a tendency for people to look down on women who are single, divorced, or have been married more than once.

In Nepalese society, there is a lot of pressure on women to bear children, especially sons. Even though this has been lessened to some extent by the revolution, there is still pressure on women to have at least one child. And women are still expected to take most if not full responsibility for taking care of the children.

Speaking about some women who have joined the revolution, Parvati writes:

'With the birth of every child she sinks deeper into domestic slavery. In fact, many women who have been active in People's War in Nepal are found to complain that having babies is like being under disciplinary action because they are cut off from the Party activities for a long period. In this way many bright aspiring communist women are at risk of being lost in oblivion, even after getting married to the comrades of their choice. This is specially so in white dominated areas [areas still dominated by the local traditional elite] where women seldom get support from the mass as well as from the Party to sustain themselves in their reproductive years.'

Parvati also raises the problem of 'conservatism' in the Party which leads to 'relegating women cadres to only women-related work, thereby robbing them of the chance to develop in party policy matters and other fields.' She points out how, spontaneously, women's issues may get talked about but not implemented. She says:

'Often it is seen that the party does not actively intervene in the existing traditional division of labor between men and women whereby men take to mental work while women are left to do physical labor. This is also manifested in taking men and women as absolute equals by not being sensitive to women's special condition and their special needs. This becomes all the more apparent when women are menstruating or are in the reproductive period.'

Feudal family relations and obligations also exert their influence on how women look at themselves and how men and women relate to each other in the Party and in the PLA. Some women may view marriage and motherhood as a break in their political/military career. Parvati points out that women cadres sometimes 'follow the directives of the Party blindly without questioning, just as traditional women have been following their fathers when unmarried, and their husbands when married, and their sons when widowed.' She says this sometimes results in things like unplanned pregnancies and women following their husband's political line blindly.

In terms of men in the revolutionary movement, Parvati points out that while women have problems of asserting themselves, men have problems with 'relinquishing the privileged position bestowed on them by the patriarchal structure.' For example, men may formally accept women's leadership – but not really or fully accept and respect women leaders. She says men are sometimes 'impatient with women's mistakes and general lack of skill in fields from which women have been excluded' and in general may not pay much attention to women's issues. There is also the tendency for men to revert to the traditional division of labor in which men do the 'mental work' while women are relegated to menial tasks.

These are all real problems that have arisen in the course of trying to develop women leaders in the People's War. At the same time, the CPN (Maoist) has made progress on this front.

At the time of my trip, in 1999, many squads and platoons had women members. But there were very few women in leadership positions in the People's Army or the Party. Now, Parvati notes, there have been real advances in developing women leaders and recruiting women into the ranks of the revolution: As of 2003, there are several women in the Central Committee of the Party, dozens of women at the regional level, hundreds in the district levels, and several thousand in the area and cell levels in the Party. In the People's Liberation Army there are many women commanders, vice-commanders in different sections within the brigade, platoons, squads, and militia. And in the United Revolutionary People's Council, the embryonic central government of the areas under Maoist control, there are 4 women out of 37 members. Women's participation in all levels of the People's Councils has also been made mandatory. In the Western Region of Nepal alone, there are 1,500 women's units. The total membership in the women's mass organization is 600,000. In the military field, there are ten women section commanders in the main force, two women platoon commanders in the secondary force, and several militia commanders in the basic force. And the team commander of the health section of a battalion force is a woman.

18
Magar Liberation

We travel a lot at night. But in more secure areas we are able to journey during the day. Most of the time we go for hours and hours, on and off well-worn trails, without running into other people. But once in a while, we find ourselves sharing the path with local villagers going about their daily routine.

One day we hit the road early in the morning, going uphill. For a couple of hours we wind our way on this trail by ourselves. But then we begin to encounter a steady stream of peasants going downhill. Men and women carrying huge loads on their backs nod at us as we cross paths. Others pass in silence with their heads down, eyes focused on the ground before them. There are even small children trudging down the road, carrying heavy plastic jugs.

I ask Pravat what they are carrying. But he doesn't know, so he asks one of the people traveling with us who is familiar with the local area. We find out that many peasants around here collect pine tar from trees high up in these mountains. Then they lug it down the hill and sell it to a middleman, who sells it to some business. I can see that it is tremendously hard, back-breaking work – another example of how peasants in Nepal must constantly search for ways to supplement their income because they cannot survive just by farming the land.

We have been in Rukum now for about a week and the conditions of our travel have gotten more dangerous. It is almost the first election day of the two-phase election, and the districts picked for the first round of voting are all the areas where fighting between the guerrillas and the police has been the most intense. So thousands of police have flooded into the Western Region, especially into Rolpa and Rukum. Military encounters between the People's Army and police have really escalated over the last month.

Last night we got to a shelter around 9:00 pm and after eating a meal, it seemed like we were settled in for the night. But then, just after midnight, we suddenly had to leave and travel for a couple more hours over rugged terrain. Later, I found out that we had been only a couple of miles away from one of the areas being set up as a polling station and the guerrillas had received information that a lot of police were patrolling the area. So we had to get away quickly under cover of night.

One of the aspects of the People's War I'm very anxious to learn more about is the relationship between the 'New Democratic Revolution' and the struggle of the different oppressed nationalities in Nepal.

Ethnically and culturally, Nepal is a very diverse country with some 30 major ethnic groups and 100 different languages. The different nationalities in Nepal each have their own territorial base, culture, and language.

The revolutionaries consider the national question one of the most important components of the New Democratic Revolution in Nepal. And during my visits to other parts of Nepal I had already gotten a glimpse of how the struggle of different ethnic groups is fueling the People's War.

In Kathmandu I had talked with the chairman of Newa Khala, the organization which had led the *bandh* (strike) around the demands of the Newar people. Then in the east, I interviewed guerrillas who were members of the Tamang oppressed nationality. They said one of the main reasons they had joined the People's War is because they want to fight against the whole way the government oppresses ethnic groups that are considered 'lower caste.'

One 22-year-old man told me: 'I come from a simple farming family, from a lower caste. I have been oppressed in two ways – oppressed generally as a poor peasant and oppressed racially. When the People's War started, I became excited and thought this was the way to be free from the reactionary Hindu state.'

Another member of this squad said: 'The main reason I joined the People's Army is because there is discrimination by Hindu chauvinism, by the reactionary ruling class against the indigenous people. I belong to the Tamang people, one of the oppressed nationalities in Nepal. This group doesn't have any opportunities in the government and the ruling sectors, and we have been oppressed by the reactionary Hindu state. I came to understand that, in order to get free from this kind of oppression, we cannot do this without picking up the gun, because the reactionary state power rules over us with the gun. This is why I joined the People's Army.'

An older guerrilla, a 40-year-old man from Bethan, said: 'The main reason I joined the People's Army was not only economic repression, but as an indigenous people we can't speak our language, read our mother tongue, and we are repressed by the Hindu government. So now I have great hope and determination that we will be able to establish a New Democratic system that is for equality and will wipe out all the discrimination that is being done by reactionaries.'

Here in the west, the Magar people are the largest ethnic minority group. And in Rolpa and Rukum, they make up the majority of the forces fighting for the revolution. In fact, the strong unity between the struggle of the oppressed Magar people and the revolution is one of the reasons these districts have become strongholds of the People's War.

The Magar people here look distinctly different than other people in Nepal – they look more like Chinese people and some of them do not speak Nepali. Some of the families I interview only speak their native language, Kham. So I have two translators – one to translate from Kham into Nepali and the second to translate from Nepali into English.

Nepal was established in the second half of the eighteenth century through the forcible annexation of nearly 60 different tribal and ancient states, scattered along the mountainous terrain south of the Himalayas. Before this, the land which now makes up Nepal – about 500 miles by 100 miles in size – had been a mixture of migrating groups of people from the Indian plains and the Tibetan plateau for about 3,000 years.

People with Tibeto-Burman or East Asian roots inhabited the eastern and central parts, while people with mostly Indonesian-Aryan roots occupied the western part of the countryside. Then, after the twelfth century, in the wake of the Mogul invasions of India, there was a great influx of Hindu migrants into the areas that would eventually become Nepal.

By the fourteenth century, petty feudal kingdoms had been established through the gradual assimilation of indigenous tribal communities in most of the central and western hill regions. Out of this situation people with Indonesian-Aryan roots, particularly the upper-caste strata of Brahmins and Chhetris, came to dominate other ethnic groups. From this time on, the Khas nationality, the Hindu religion, and the Nepalese language have been imposed throughout the country. And for centuries, the Nepalese ruling class has exercised discrimination, exploitation, and oppression against other religions, languages and nationalities.

During my travel in the west I meet with Narenda Buda, a Central Committee member of the Nepal Magar Association and a member of the All Nepal Nationality Association. He tells me:

'In Rolpa, 80 percent of the people are Magar. In Rukum it's about 65 percent, in Jajarkot, about 40 percent. Sallyan is about 40 percent Magar. In the west as a whole it's about 40 to 50 percent Magar. Almost all the Magar here are poor peasants. They mainly practice natural shamanism with local priests [*Jhankri*]. But religion is not that strongly practiced among the Magar people. They have been dominated by the Hindu religion, which has suppressed the Magars' indigenous religion and enforced the practice of Hinduism. For example, Magars used to eat beef but the government banned this.

'The Magars have demanded to study in their own languages – Kham, Magar, and Kaike. But the government won't allow this. Most Magars are bilingual, speaking their native language and Nepali. The government teaches Sanskrit, which is a dead language, in the schools; and this makes it even harder for Magars. Most of them fail in this subject and there is strong opposition to learning this language. There have been processions, the burning of Sanskrit books, and boycotts of Sanskrit exams. Still the government doesn't listen to this demand.'

Narenda explains that there are many supporters of the People's War in the Nepal Magar Association. But he says there are also many other political forces in the group, including Magars who are associated with groups hostile to the Maoists. He says:

'The aim of the Nepal Magar Association is to promote and advance the culture and language and improve economic standards. It is mainly a social and cultural group. There are indigenous Magars who have a long cultural history and live in particular areas. Some indigenous Magars have been assimilated and have lost much of their culture. Some ethnics [or tribes] are not indigenous and are dominated by other castes and live in primitive social and economic conditions.'

On the other hand, the Magarat Liberation Front is a much more political organization that openly promotes the People's War. Narenda says:

'The Magarat Liberation Front agrees with the Maoist position and demands to participate in the revolutionary united front. MLF demands self-determination from the reactionary government that oppressed them economically, politically and socially. If a new democratic government comes to power and oppresses the Magars, this demand would remain. But if a New Democratic government gives Magars equality and opportunity in every aspect, they will not demand a separate autonomy, but will participate in the New Democratic government. They are hopeful that the Maoists will give the Magar people equality in all aspects of society and will not be compelled to make a separate nation.'

When I interview Anuman, a Central Committee member of the Nepal Magarat Liberation Front, he tells me:

'In Magar society there is a primitive kind of communistic/collective way of living. Some farms are worked collectively and at harvest time things are divided up. In the forest the people look after the grassland as communal property to be used by everyone. People live very closely together, sometimes many generations in one house. And among the Magar there is more equality between women and men. When the MLF goes to the masses we first give priority to the indigenous culture and language and problems of social and economic conditions, exposing how the Magars are dominated by Hinduism. We have to prevent our culture and language from disappearing.'

The Communist Party of Nepal (Maoist) has a policy on self-determination and autonomy for the oppressed nationalities in Nepal. They say they want to establish a New Democratic state, with the joint participation of all nationalities, the ending of all forms of oppression and exploitation based on nationality, language, religion, etc., and equal treatment and opportunity for all ethnic groups and languages.[1]

Anuman says, 'We say we should be united to seize power and put power in the hands of the lower caste people. And in our area we should manage all the different aspects of society – construction, schools, etc. From a social point of view, Magars are very backward and high castes ignore us. National politics are in the hands of other castes and give no power to the Magars. We are oppressed by Hindu chauvinism, but our view is not to establish Magar chauvinism. We are for the equality of all castes and respect all religions and cultures. Our view is for equal rights of all human beings in all aspects. But we are not getting equality now so we should unite and get equality.'

* * *

After we finish our meeting I think about how centuries of feudalism have created an extremely hierarchical society in Nepal. Certain castes are considered to have 'god-given' rights to rule over everyone, while other castes and ethnic groups are deemed 'untouchable,' fit only to do the dirty jobs in society. This kind of discrimination and national oppression here has its own history and uniqueness. But at the same time I can't help but think about how it is not all that different from how things are throughout the world.

In traditional Nepalese society it is considered sacrilegious and totally unacceptable to marry outside your caste or ethnic group. And upper castes are certainly not supposed to marry someone from a lower caste. But the Maoists are trying to institute new customs and attitudes about the tradition of castes. Many revolutionaries here have introduced me to their wives or husbands, proudly saying they are an 'inter-caste' marriage. And when young guerrillas tell me they are from a poor peasant family, from a lower caste, they hold their heads up high, saying they are proud to be fighting for a society where such a hierarchy no longer exists.

19
Preparing for War in Rukum

Nights in the mountains of Nepal are usually cool at this time of the year, and sometimes pretty cold when a strong wind gets blowing. But when the sun reaches high noon and then crawls across the western sky, it is pleasantly hot.

A lot of the women here wear their long hair off their necks, out of their eyes, tied back in a bun. This seems like the ideal hairdo for hiking up and down the mountains in the heat. I too have long hair, so one day I ask Onsari, one of the women squad members, to help me fix my hair 'Nepali-style.' She runs a comb through my hair several times, then quickly twists it into a tight knot and winds an elastic band around to keep it in place.

The next morning, on my way back from a trip to the forest, I see Onsari has been waiting for me. She's sitting on a log making something and flashes me a shy smile. When I walk over she hands me a present – she has crocheted bright red yarn into a beautiful hair band for me to wear in my new 'Nepali hairstyle.'

Onsari has been traveling with me since we crossed the border into Rukum. We've been staying at this shelter now for a couple of days and I've noticed how, like other members of the People's Army, she integrates herself into the daily life of the village. Like most of the guerrillas, Onsari is from a peasant family and grew up in a village just like this one. While we've been meeting she has been hanging out with the local women, talking, and helping out with daily tasks.

Today we are leaving this village in Rukum, and Onsari is going back to another part of the district with her squad. We pack up our things and then, around 5:00 pm, everyone gathers outside for a little farewell ceremony. The local cultural team sings a song about saying goodbye and they even add some verses about me – how I've been 'braving the difficulties of going up and down the mountains.'

Actually, I've been feeling close to being defeated by these mountains in the last couple of days. But the song gets me ready for our next trek, reminding me how much people here appreciate my effort to hear and spread their story.

Everyone who is not leaving with us lines up outside and I walk down the line to say goodbye. Onsari is at the end of the line and when I get to her, I reach back and touch my hair, which is neatly tied back with the red hair band. I give her a big smile and shake her hand. I ask Pravat to tell Onsari 'thank you' and that I will always remember her. I can see that a few tears are welling up in her eyes.

We leave at 7:00 pm and reach our next shelter at around 1:00 am. It is the eve of election day and at one point, we can see on a mountain, far across the way, flames from a torchlight procession, dimly flickering in the growing darkness. We also hear the faint sound of gunfire echoing off the mountainsides – villagers demonstrating against the elections are firing into the air.

By the time we reach our shelter it is very late. But the people gathered here – high-level Party and military leaders of this district – have been waiting anxiously for us and want to have a little greeting ceremony. By the time we finish our meal of *dal bhat*, it is 2:30 am and people can see that I'm tired, so they tell me to go to sleep while they stay up to talk.

The next morning begins two days of discussion with the District Committee Secretary (DCS) of Rukum and a member of the Party's Central Committee (CC) who is in charge of the work in this main guerrilla zone.

In the years leading up to 1996, the District leadership here launched a number of campaigns to prepare people for making the huge change to waging armed struggle. They carried out what they call 'rectification' – education and struggle to improve working methods and unite people around the Party's line. The DCS tells me that in this area there weren't a lot of people who disagreed with the decision to launch armed struggle against the government, but some people were afraid and did not want to make the necessary sacrifices. The DCS tells me:

> 'Some were won over while some retreated. Some new leadership came forward ... At the mass level many campaigns were conducted. First we did mass propaganda and also different construction projects in order to spread the influence of the Party. Before this campaign about 60 percent of the people were influenced in some way by the Party. Afterwards it went to 80 percent. When more people were attracted to the Party the leadership was encouraged. Just before the Initiation, the Party leadership gave many political classes to the masses because we saw the masses as the basis for the People's War; and so if they were not politically conscious, we could not be successful in carrying out our program.'

The CC member then goes more into the different aspects of the preparation that went on in Rukum:

'In terms of political preparation, mass propaganda was given to the masses about New Democratic Revolution and Marxism-Leninism-Maoism and the necessity of protracted people's war to be able to seize power. The forms of propaganda were mass meetings, cultural campaigns, postering and walling, pamphlets, newspaper articles, and political classes. This was done at a mass level as well as among cadres and the Party leadership.

'This went on from 1992 to 1996 and accelerated in 1995. At the time there was sharp class struggle – in the villages there were actions against usurers, cheaters, and liars, who were exploiting the people. These were all sabotage-type actions, not annihilations. The Party grasped various issues to wage mass struggle against the government and there were many mass movements organized, among students, women, peasants, intellectuals.

'During the preparation the police charged about 10,000 people with different cases in Rolpa. Large numbers of people were arrested and jailed. Some were tortured and released, others were kept in jail for a long time. Many were forced to go underground.

'The government oppressed the mass leaders and cadres viciously. There was the Operation Darbot and the Operation Romeo. These two oppressive campaigns prepared the objective ground for people's war nationwide. Subjectively, the Party was prepared internally. Here in Rukum, we spread propaganda about the oppression going on in Rolpa and this created a favorable political situation to start the People's War. More people became in favor of the Maoists nationwide.'

Next the CC member talks about the kind of military and 'struggle' preparation they carried out:

'Three types of military organization were formed – fighting teams, village security teams, and volunteer teams. There was physical training and training in how to use weapons.

'The Revolutionary United Front has two functions. It is a means to conduct struggle on a central level. And on the local level, it is a means to conduct struggle as well as seize people's power. There are two types of mass organizations: those that directly support the Party and the Party line – organizations of students, women, and peasants – and those that give indirect support to the Party – like intellectuals, lawyers, and human rights groups, which cannot openly support the Party but actively work in their different fields to indirectly build support for the People's War. The mass organizations function as a web among the masses.'

In terms of 'struggle preparation,' the CC member says there was a legal and illegal struggle. Legal struggle was conducted through mass organizations, campaigns, and propaganda. At the same time, the Party was leading different kinds of guerrilla actions – raids, ambushes, commando attacks, and setting mines. He explains:

> 'Commando attack means a surprise attack by a small force against a bigger force, armed or unarmed. The purpose of a commando attack is to seize arms and annihilate and harass the enemy. The attack is very quick, the time is short.
>
> 'In terms of technical preparation, our weapons included three types – domestic, firearms, and explosives. Domestic weapons include knives, sticks, and other simple weapons. Firearms include muzzle-loader rifles and automatic guns. Explosives are grenades and mines. The policy is to use basic domestic arms and advance to firearms.
>
> 'Before the Initiation there was collection of all these weapons from the people and some were bought. The Party also gave physical training to the fighter teams and training in how to use arms. The Central Military Commission of the Party's Central Committee developed a guidebook on the general military line and educated the fighting groups and leadership groups, down to the District Committees.'

* * *

We've been talking for several hours now and decide to end for the day. So I will have to wait until tomorrow to hear the rest of the story about how the People's War got started in Rukum and how this district has become one of the strongholds of the revolution.

While we have been meeting today, the government has been trying to carry out the first phase of their elections. Someone comes into the room and says, 'We have been watching the nearby polling station all day. Come outside and have a look.'

We go out and walk to the edge of a nearby ridge. A guerrilla hands me his binoculars and points to a spot in the distance. Through the binoculars I can see a small house, with a table set up outside. A couple of police are standing around with nothing to do. According to the Party there is little support for the government in this guerrilla zone and most people are boycotting the elections. Someone who has been watching the polling station all morning says, 'Except for a couple of people, nobody has voted at this polling station all day.'

20
Starting and Sustaining People's War in Rukum

I've been in the Rolpa and Rukum countryside for several weeks now. The main highway is days of hiking away and our only real connection to what is going on in the rest of the world has been the little radios some of the guerrillas carry which can get the BBC news. Being in one of the most remote places on earth, it sometimes feels like what's going on here is really isolated from the rest of the world. But paradoxically, it really strikes me that wherever I go, people make a point of saying that what they are doing is 'part of the worldwide revolutionary struggle.'

The *Internationale* is a song sung by communists around the world. Back in the United States, I was familiar with it in English and Spanish. But here I had heard it sung in Nepalese for the first time and want to capture this on tape so people back in the United States can hear it.

People tell me that the District Secretary of Rukum, whom I have been meeting with, is a man of many talents – in addition to being a political leader, he also composes and sings revolutionary songs. When I hear about this, I immediately make a request. I ask him if he will let me make a tape recording of him singing the *Internationale*.

So after our evening meal, the DCS and his wife, who is also known for her singing, put on a little performance for me. After singing several revolutionary songs to warm up, they sing the *Internationale* so I can tape it. Then, to my surprise, there is, in turn, another request. The DCS pulls out his own tape recorder and asks if I will sing the *Internationale* in English, so he can record it.

Unlike the DCS and his wife, I've never been known as a good singer (to say the least). So at first I panic at the thought of my slightly off-key rendition of the *Internationale* being circulated here in Nepal. But then I manage to swallow my pride and sing into the microphone. The people in the room can't understand the English words. But they smile broadly as they recognize the tune.

* * *

The next day I continue my discussions with the DCS and the Central Committee member in charge of this zone. After a year of intense preparation, armed struggle started across the country on February 13, 1996. The DCS describes what happened in Rukum during this First Plan of the Initiation:

> 'The military action we were assigned was a raid on a police post and it was successfully carried out on the day of Initiation. The aim was to capture weapons, but there were no arms at this police post. The cadres captured the seven police and explained the political line behind the action to them. Official papers were seized. On the same day, the same detachment carried out a sabotage of a shop owned by a usurer, a man who exploited the people.
>
> 'Within 15 days of the Initiation, many military actions were carried out. There were 500 actions during this time. Most were sabotage actions aimed at usurers and liars, who cheat the people out of land and money, and spies.
>
> 'At the same time, the masses were involved in many different mass actions, like processions, cultural programs, meetings, etc. The Party leadership and cadres were encouraged after the Initiation. On the other hand, the enemy, the government, was confused about how to react.
>
> 'After 15 days, the government counterattacked very strongly. The government's home minister came to this district, stayed in the district headquarters, and gathered together with local officials and reactionaries to plan how to attack the People's War. They planned to kill many Maoists to try and stop the People's War. Six people were killed in one place, in Pipal VDC in Rukum, and there were all kinds of raids, arrests, rapes, etc.
>
> 'Only a few Party cadres and leaders were underground before the Initiation. Then with the Initiation, 90 percent of Party leaders and cadres went underground, and many sympathizers also went underground.'

The CC member talks about the achievements during the Initiation in this whole zone, which is a larger area than just the Rukum District. He says:

> 'There were three categories of work in the Initiation – guerrilla actions, propaganda, and sabotage. On the first day of the Initiation in this zone, there were two guerrilla actions – the raid on the Holeri police post in

Rolpa and the raid on the Rari police post in Rukum. There were three main sabotages and other smaller sabotages and many propaganda actions. In the first 15 days, the total number of military actions in this zone was 1,800. The main slogan for these actions was: "Destroy the Monarchist Parliamentary System and Establish a New Democratic Republic of Nepal."

'Before the Initiation, actions against the enemy were classified into four types – actions against: 1) the reactionary government; 2) feudal representatives; 3) comprador capitalists; and 4) imperialist institutions. In the Initiation the fighter groups targeted all four types of enemies.'

The Party's Second Strategic Plan was started eight months after the Initiation, in October 1996. It was carried out under the slogan, 'Develop guerrilla warfare in a planned manner so as to prepare the ground to convert specific areas into Guerrilla Zones.'

The CC member says:

'In the Second Plan, more and more cadres came into the Party. This was one achievement of the Second Plan. The Party established that the main form of organization is military and the main form of struggle is war. Fighter groups were promoted to squads. In military terms there was quantitative and qualitative growth. The situation demanded the struggle to advance.

'The people accepted the People's War and became involved. The influence of liars, cheaters, and usurers in the village decreased. Some of them moved away to the district headquarters or the capital, and those who stayed in the villages were not as active as before. During the Second Plan we started annihilations and raids, and ambushes continued.'

The DCS says:

'At the time of the Initiation there were three types of military groups: 1) fighter teams; 2) village security teams; and 3) volunteer teams. The fighter teams carried out the raid on the police post and were the advanced team at the time. In the Second Plan squads were formed from the fighter teams. Fighter teams were then dissolved, but the other two types of military groups remained.'

He then describes what effect the development of armed struggle had on the kinds of people that came forward to join the Party:

'In the three years of People's War, the Party's composition here has changed. By the Second Plan, most of the Party cadre and leadership were men and women from poor, peasant backgrounds. There were less intellectuals and students. Now most people in the Party here are young, 15 to 35 years old, with the majority 18 to 30 years old.

'Before the Initiation, the membership of the Party decreased through the rectification campaign. But in three years of People's War, Party membership here, which includes the People's Army, has doubled.

'The criteria for membership is stricter now than before the Initiation. Before, the Party mainly conducted legal struggle, and more people could be involved in this type of struggle, and it was easier. A lot of people who worked with the Party became members and at the time the risks were not as great. But now is a time of war. The Party is mainly involved in illegal activities and needs more dedicated cadres who are willing to sacrifice and keep strict secrecy. Today, the criteria for membership is higher in terms of ideological and political unity and willingness to make whatever physical sacrifices are necessary for the People's War and the Party.'

In August 1997, the Party launched the Third Strategic Plan to 'Develop Guerrilla Warfare to New Heights.' This entailed raising the military ability of the revolutionary forces both qualitatively and quantitatively. According to an editorial in the Party's journal, *The Worker*, this meant 'the creating of a base for local organs of political power and raising the political, organizational and technical level of the people's guerrilla army so as to be able to contend with the rival army in the prospective guerrilla zones.'

The DCS tells me:

'By the start of the Third Strategic Plan, there were many squads, each with a minimum of five and a maximum of nine people in each. And there were about 60 to 65 village security teams and many volunteer teams.

'The squads have three types of weapons – knives, muzzle rifles, and 12-bore rifles (basic, secondary, and leading). Village security teams have knives and volunteer groups have domestic arms (knives, sickles, sticks, etc.).'

The CC member tells me that military actions increased during the Third Plan and he points to the significance of the boycott led by the Party against the local elections:

'In this zone there was a complete political vacuum in about a third of the VDCs [local political areas designated by the government], where there were no VDC officials. In another third of the VDCs, there was a partial power vacuum. The important achievement of this period was the election boycott and the new thing that developed was this power vacuum. This was the result of the political movement and military actions. With this development the Party introduced the plan to develop base areas.'

The Fourth Strategic Plan started in October 1998 under the slogan 'Go forward to establish base areas.' During this period, the People's Army in Rukum was able to achieve a higher level of military organization. While maintaining the squads, the first platoon was formed with about 24 members. The DCS says:

'In the platoon the leading [main] weapons are automatic or semi-automatic rifles, with a minimum of one for each platoon. The secondary weapons are 303 bolt-action rifles. And the basic weapons are 12-bore rifles. In the squads, the main weapon is 303 bolt-action rifles, secondary is 12-bore rifles; and basic is muzzle rifles. Platoon political commissars and platoon commanders have revolvers or automatic pistols.'

With the Fourth Strategic Plan, many more militias were formed in Rukum and the DCS tells me this includes a number of all-women militias. The squads are about 30 percent women and there are some women who are vice-commanders of squads. But at this point, there are not any women commanders.

The CC member explains that in the Fourth Plan, four categories of areas were established: 1) prospective base areas; 2) prospective guerrilla zones; 3) guerrilla action zones; and 4) propaganda zones. And there are now three types of military forces: 1) the main force, platoons; 2) the secondary force, squads; and 3) the basic force, militias, and volunteer groups. He says:

'The platoons and squads are led directly by the Party. The militias are led by the revolutionary united front. The platoons are under the regional bureau; squads are under district leadership.

'The platoons carry out guerrilla actions – primarily raids of police posts with the main aim to drive the enemy away and get weapons, and secondarily ambushes – setting mines and commando attacks. The squads mainly carry out ambushes and secondarily carry out minings, commando attacks, and raids. The task of the militias is first of all sabotage and secondarily propaganda.

'All forces are allowed to carry out annihilations. Higher types of annihilations, like the killing of bureaucrats and political leaders, are done by platoons. Squads and militias target usurers and other lower level class enemies.

'All military forces have five functions: 1) political work (mass political education); 2) organization work (developing mass organizations); 3) production and construction (growing food grains and solving people's problems); 4) propaganda (putting slogans on walls, postering, and processions, including armed marches with torches); and 5) fighting, military actions, which is the main function of the military organizations.

'The main force moves to and from the prospective base areas to the guerrilla action zone. The secondary force is also in the prospective base areas and can combine with the platoon for actions. There are also squads in the prospective guerrilla zones.

'Platoons cannot stay in the guerrilla action zone very long and there are only a few squads in this zone. The main force is very mobile while the squads are fixed in their areas (prospective base areas or prospective guerrilla zones).

'Squads and militias are in the propaganda zone – they centralize for actions and then decentralize. The main aim of these actions is to carry out attacks in the urban areas.

'In the prospective base areas our aim is to seize power and establish a permanent base area. Now our control in these areas is temporary – sometimes seized by the people, sometimes seized by the enemy.

'In the prospective guerrilla zones, the aim now is to establish temporary base areas. In the guerrilla action areas, the aim is to demoralize the enemy, extend the ground for warfare, and seize arms.

'In the propaganda zones, in the urban areas, the aim is to attack the enemy there and prepare the ground for eventual insurrection.

'The function of the Revolutionary United Front is to 1) seize and practice people's power in the [prospective] base areas; 2) seize and practice people's power and struggle in the guerrilla zones; and 3) carry out struggle in guerrilla action zones and propaganda zones.

'Since the Initiation, in three years in this zone, the main military actions have been: seven raids, 28 ambushes, and minings. And there have been more than 50 annihilations, 49 police have been killed.

'In short the achievements in three years are: 1) from the political point of view, local power has been seized and practiced by the people;

2) from the military point of view, three types of forces have been formed and there has been the beginning of mobile warfare; 3) from the cultural point of view, we are eliminating feudal culture and introducing a new revolutionary culture; and 4) from the economic point of view, we are carrying out collective and cooperative production and volunteer production teams are helping martyr families and others who are not at home. Interest rates are fixed by people's power. There are fixed prices for goods. Many capitalist and imperialist institutions have been driven away. There is the start of cooperative financial funds. We are beginning to establish economic self sufficiency and we are carrying out production and construction – building paths, schools, playgrounds, martyr monuments and gates, bridges, public toilets, etc.

'In the last three years we have gone to the base class, the oppressed, and we have recruited many women into the different levels of supporting the People's War.

'There is one problem and that is: we have not been able to advance the military ability that is needed. In the prospective base area zones there are sufficient cadres but in the prospective guerrilla zones there are not enough so some cadres from the prospective base area have to be transferred to the prospective guerrilla zones. They go there temporarily to help build the work and then may return. But there is a lack of cadres, quantitatively and qualitatively, in terms of leadership and military ability.'

The DCS adds:

'We are clear on the political and ideological line in the party and among the masses. But one contradiction now is that the platoons are not as qualitatively developed as what is needed, in terms of military skills. And there is a big problem of not enough advanced weapons. We cannot fulfill the criteria we have set for the platoons and squads. If we have more weapons, we can give more training and increase the number of platoons and to establish a base area, we have to develop the military skills of the platoons and squads and provide arms according to our criteria.

'Within three years, we have learned that no one is born brave, but can become brave in practice. When we practice warfare we learn more lessons about war and it makes us mentally prepared to fight. Now, mentally we are prepared to fight, and we are ideologically and politically clear. But we have a crisis of arms – we don't have the necessary weapons. We are advancing our skills for warfare. We have learned that even though we have inferior weapons, if we are mentally and ideologically and politically prepared, we can defeat the enemy. At the same time, in practice, we need sophisticated arms to get victory in the battlefield. So now we are concentrating on getting/capturing more sophisticated weapons.'

21
Camping with the People's Army

The moon here in western Nepal is rising after midnight now, so it is better to travel early in the evening, before it's really dark, or in the early morning hours. Tonight we leave just after sunset and travel for about two hours, then stop at a peasant's house for a late meal. We sleep for a few hours, and by 4:00 am we are back on the trail.

We reach our next shelter around 8:00 am, and we are able to rest there the whole day. Then, in the evening, we leave for a hard, two-hour steep climb. We make our way through thick brush, some of the time in total darkness. The guerrillas have to use their flashlights, flicking them on for a few seconds in order to see where we're going. Then it's back to groping our way in the dark.

It is late at night when we reach a campsite where many other people are gathering. The squad members fix up a little sleeping area under some trees and we quickly settle down to sleep. I am tired but excited and I can't fall asleep right away.

The Central Committee of the CPN (Maoist) has organized a special group of guerrillas to carry out military actions during the election boycott campaign. This 'special task force' is bigger than the platoons, which have about 27 members. With more than 50 members, it is more like a company, which is the next higher military unit. During the night, members of this task force will be arriving at this camp and tomorrow I will meet them.

I wake up very early in the morning and the camp is already buzzing with activity. I can now see that we are at the top of a big mountain, on a plateau. There are dozens of guerrillas walking around, dressed in army fatigues. A tall pole has been put up in a large open area and, at the top, a red hammer-and-sickle flag flutters in the early morning breeze. A group of people are busy putting up banners and setting up for a program.

Some task force members come over and introduce themselves. They show me some of the weapons they have recently captured. There is a shiny Smith & Wesson .38 revolver, taken from a police inspector killed in a recent mining action. There is a walkie-talkie which the guerrillas have been using to listen in on police conversations. And there is a high-powered rifle with a telescope, which was captured from a tourist who was hunting. In fact, I had read about this incident in the *Kathmandu Post*, about how the guerrillas had demanded that the tourist turn over his weapon before letting him go unharmed.

One man is walking around wearing a 'new hat.'[1] He explains to me that just the other day, it was taken from an election official, after an ambush/explosion set off by the task force. He guesses that the official might not be alive any more because the hat has two big holes in the front from shrapnel.

Twenty-four-year-old Platoon Commander Sundar[2] is one of the task force members who come over to talk. He has a very young, boyish face in stark contrast to the seriousness of his military responsibilities. He has come to this camp fresh from recent military actions leading up to the elections. When I interview him he starts by telling me a little about the history of the platoon he commands:

> 'Our platoon is under the command of the regional bureau. There are 24 members including the political commissar and platoon commander and platoon vice-commander. The platoon is also divided into sections, with section commanders. The members of our platoon are from 18 to 32 years old, with the majority 20 to 24 years old.
>
> 'Our platoon was formed in September 1998. We centralized the work of the platoon for a major campaign and at that time the military work was to do sabotage-type raids.
>
> 'The platoon chose the target for the sabotage/raid and went to the police post. But the police knew of our plan. The platoon was taking shelter and the police went there. The platoon left and met the police who were on their way. There was an encounter, with shooting back and forth. A District Committee member was killed by the police in this encounter and some police were possibly injured. The police ran away and the platoon left safely.
>
> 'In the local elections in 1998 the platoon decentralized to give political classes to the masses to build the campaign to boycott the elections. Also, we prepared to ambush the police during this time, but the police did not come into our area.
>
> 'During this time, one of the platoon section vice-commanders was killed by the police. He went to one village to prepare an ambush, he wanted to survey the area. He was taking shelter in a house and the police found out about his presence and came there and killed him. This was martyr Chain Buda, "Andhi."
>
> 'The people of this area were frightened after this and the platoon could not get shelter and carry out a military action there right away.

The platoon decentralized and had political discussions with people in the area and raised people's political consciousness. The platoon also made a monument to martyrs in the area – a martyr platform under a tree and a small pond for the peasants' cattle.'

Next, Commander Sundar tells me about what the special task force has accomplished:

'The task force is temporary, formed to carry out a particular action or campaign. Our platoon sent a number of members to such a task force. The main job of such task forces is raids and secondarily ambushes.

'Recently, the task force carried out two raids – on the Dang Chiraghat police post and on the Rolpa Jelwang police post. In Chiraghat there was a one-and-a-half-hour encounter. Seven policemen and one task force member were killed. The task force got six rifles and one revolver (made in China), 224 bullets and other equipment from the police post. All the police who were at the post were killed.

'In Jelwang there are two police posts, one regular and one commando. The task force raided the commando post. There were 16 police there. There are other police posts nearby, within a half hour. So the task force made a plan to raid the regular police post at the same time, to prevent them from coming to the commando police post.

'We also prepared to ambush a third police force if it came from the other nearby police post. The task force went to the police post according to plan. The action involved other platoon and squad members in addition to the task force. The task force raided the commando post. First, we captured the sentry and killed him. The police were alerted to the attack and started shooting. The encounter lasted one hour and 15 minutes. The task force threw a grenade at the police post and burned it down. It is unclear how many police were able to escape. This happened the first week of April 1999. We captured one rifle from the sentry and one bulletproof jacket and 23 bullets.

'On April 17, 1999, our platoon set up a big ambush in Pokhara in Rukum. Five police were killed. The police of this post had killed many people in Rukum and this action was to get revenge and was part of the election boycott campaign.

'After these actions the task force was decentralized. Now a small section of the task force is giving priority to doing ambushes.'

I ask Sundar how the guerrillas in his platoon are given military and political training and he says:

'Military training is given by the Party's Central Committee and the platoon provides other training as necessary. There are two types of military training: one is physical training – jumping, running, crawling, etc.; the other is technological training – how to use weapons. When the Party makes a plan it gives political classes to the platoon to make clear the aim of the plan and how to fulfill it. The platoon is also given political classes in Marxism-Leninism-Maoism. There is a military department of the Central Committee that gives classes to the platoon members, which includes the study of military strategy.

'Politically, the People's Army has to do mass propaganda work among the people to counter the lies of the enemy. And militarily it has to break out of the encirclement by the enemy. We have synthesized bourgeois and revolutionary warfare. We can develop our own military warfare, learning from bourgeois and revolutionary theory. There is a history of different kinds of warfare, and the Party leadership gives us this experience. We also use our own experience, but it is not enough. We are following the model of protracted war waged by Mao in China.'

There is a huge shortage of doctors in Nepal and most doctors are based in the cities. So in the countryside, it is extremely difficult for the people to get medical care. In Kathmandu I had talked with a Party supporter who does political work among doctors. He told me there are many doctors who support the People's War, but at this point, it is still very difficult to get them to leave the city and go to the countryside to work in the guerrilla zones. So while the Party continues to struggle to get professional medical care for people in the war zones and wounded guerrillas, it also has a program to train people so that they can provide basic first aid and health care.

Twenty-five-year-old Gaule is in the People's Army, serving as the medic for this task force. In the afternoon I sit down to interview him and he tells me how he is learning and practicing medicine in order to serve this revolution. He says:

'I passed the Clinical Medical Auxiliary Exam to be an Auxiliary Health Worker. It is a one-year medical course. I completed the Intermediate Examination in education [teaching] and started bachelor-level work but didn't complete this. I left to join the Party's work. I worked in a government sub-health post in a village up until four months ago and then became a full-timer in the revolution.

'The Party has different levels of health workers. Some have knowledge of medicine, others have none to start with. Recruits are given basic health training by those who have some knowledge. I was trained by doctors in the city who are sympathizers – a one-month course along with others like me. This was the first phase of training. There was training in how to treat bullet wounds in limbs and fractures and diseases like diarrhea, dysentery, influenza, and typhoid. The aim of the program is to develop doctors at a bachelor of medicine degree level in different phases. The Party has completed the first phase of this.

'At this point, regular doctors are not coming to the battlefield, so we have to produce doctors from the battlefield. In the battlefield, for the guerrillas who are injured, I can provide first aid, like for bullet wounds and other injuries. And among the people I give free services, like treating various diseases. This is important because villagers are very poor. That is why they have so many health problems, so many diseases. These services also help make the People's Army popular among the people. As a member of the People's Army I take political education classes, I get physical training, and have general knowledge in weapons.

'To those doctors who are sympathizers, I say it is necessary at this time to support the People's War and they should come to the villages to help the people. All physical and mental work is guided by ideology and we need to give more political education to doctors so they will be prepared to come to the battlefield.

'Before I came the guerrillas had to treat their own bullet wounds and fractures. They had to operate to take a bullet out by themselves. Sometimes they captured a doctor and demanded that he treat the wounded. There are also some sympathetic doctors who will agree to help on an irregular basis. And even some doctors who are sympathetic to other political parties are willing to help and will charge only 50 percent of their fees.'

Mid-morning people gather for a formal welcome ceremony held under the red flag. The task force members are all lined up in formation as we walk to the front of the area that has been set up for the program. There is a minute's silence for people who have died in the war and the singing of the *Internationale*.

I am introduced to the commander of the task force and he takes me down the lines of guerrillas and introduces them to me, one by one. Each one tells me his name and gives me a '*lal salaam*,' fist in the air and a handshake. Several of them make a special effort to greet me with a few words of English.

People think it is not a good idea to have such a big gathering for very long on the top of this mountain where it could be seen by a police helicopter flying by. So the program is short.

About 100 people have gathered here at this camp and it is a big job to feed everyone. A group of people have been keeping a steady fire going, providing tea for everyone first thing in the morning, and cooking up big pots of food. The People's Army, like most peasants in the countryside, can rarely afford to eat meat. But on this occasion, a buffalo has been slaughtered to feed everyone at this camp for a couple of days. At lunchtime we go down to the 'communal kitchen,' which is a short walk down the mountain.

Everything is well organized. Some people are dishing out the food. Others are eating in shifts, sitting on the side of the mountain. And there are also some people washing dishes. There is discipline as well as easygoing camaraderie here in this camp, which gives me a sense of the kind of fighting spirit that has been forged among this special task force.

After our mid-day meal, some relatives of people who have been killed come to be interviewed. One of them is 23-year-old Chandra Bahadur Khadka. His 54-year-old father, Madhu Khadka, was killed in 1998. Chandra tells me:

> 'I came home after being away for two months and stayed in the house one night. I was leaving the area again so I came home to say goodbye to my family. We talked at night for a long time about politics and the People's War. At 11 pm the police knocked on the door. I knew that it was the police. My father opened the door and one of the police ordered another one to shoot. They shot my father. First the bullet passed through his arm and my father said, "You are all dogs of the reactionary government. Why are you killing me? What's my crime?" The police fired a second bullet in his chest and my father fell to the ground.
>
> 'I was upstairs and trying to figure out how I could break out of the ring of police. I climbed onto the roof from inside. The police saw me and shot me in my leg and throat. I roared the slogan, "Long live CPN (Maoist). Long live the People's War!" After this the police stopped shooting and I jumped down. The police ran away – there were 24 police in all – and I was able to escape. In the morning the police went back to my family's house and took away my father's body. I went to a safe place and the Party helped me get treatment for my wounds.'

Sixty-four-year-old Hari Prasad has also come to talk. His son, Daulat Ram Gharti, was a squad commander in the People's Army when he was killed in 1997. Daulat's brother, Chadga Bahadur Gharti, and his sister, Tulasa Gharti, also come to tell their story. Hari Prasad tells me:

> 'When Daulat was 15 years old he started working in the revolutionary student organization. He joined the Party when he was 18 years old. After eight years he became a full-timer. In 1989 he was district chairman of the All Nepal Youth Association (Rukum). In 1994 he became a Party District Committee member. He was Alternate Secretariat of the District Committee. After the Initiation he became a squad senior commander. In 1997 he participated in a raid on a police post on the border of Rukum and Dolpa and was killed there. He was a very sociable man and popular among the masses. He was very militant, from the military point of view. He always said that the success of our actions depends on our pre-planning and survey of the target. He was very dedicated to the Party and worked hard to follow the Party line and guidance. He was very clear in his ideological and political line. And he had a lot of organizational experience.
>
> 'After my son's death the People's War has continued to advance. My son lost his life, but thousands of other sons have been involved in the People's War and are sacrificing to advance the war. My message to other martyr families is that even though we have lost a member of our family to liberate the oppressed people of the world, we are not sad, we always look forward to take revenge and to advance the People's War. We should be proud to be families of martyrs. I am ready, if the police shoot me. When you are born into the world, it is certain you will die some day. But a death for the people and the nation is very glorious. I hear the rumor that the police are going to kill the families of martyrs. I don't believe this. But if they do, there will be thousands of other families to go forward. My son was killed on the battlefield, but other sons have not retreated and are going forward.'

* * *

I had requested a photo session with the special task force, and some people have spent the day making red bandanas for all the guerrillas so that their identities will be protected. It takes quite a while for them to rustle up the material and cut the cloth into pieces, and by the late afternoon I'm starting to worry that the sun will be too low in the sky for good photos. By the time the task force is ready I calculate that I've got about 45 minutes of good light left, so I have to work quickly.

The task force goes through their exercises, one by one, and I snap away.[3] At one point they disappear into the trees and then reappear, camouflaged with tree foliage and paint on their faces. One guy comes marching in with a large tree branch strapped to his back and everyone starts laughing. The guerrillas seem relaxed as they show off their skills for the camera. But there is a serious edge to these drills, which are meant to prepare them for battle.

Before coming to Nepal, I had seen very few photos of this People's Army so I have been eager to catch the image of these guerrillas on film. I am hoping that the photos of this special task force will help capture the reality of this revolution so that people around the world for the first time can see the 'face of this People's War.'

22
Red Salute in the West

We've been traveling in Rolpa, Rukum, and Sallyan for almost a month and now we are heading back to the area where we started our journey in western Nepal. Today we're going through some beautifully terraced land that is greener than many of the other areas we've been in. Much of the time we're walking through a wide valley, so the going is easier, without so much climbing up and down. Along the way, we come across a large stone monument that has been built to the 'revolutionary martyrs.' It has been defaced by the police.

Nearby, some villagers come out to greet us and take us over to see a small hut that now has some special meaning to the people in this village. I am told that the police regularly come through and harass the people. A while ago, some guerrillas came and booby-trapped a bench outside this shack. They told people in the village to stay away from the bench and for about a month, everyone, even the small children, knew not to go near it. Then finally, some police came through and the peasants held their breath, waiting for them to sit down to rest on the bench. The villagers tell me that they had hoped that a whole bunch of the police would be hit by the mine. But when it went off, only two cops were sitting on the bench and only one was killed, the other injured. The group of villagers asks me to take a photo of them as they pose in front of the hut.

By mid-morning we reach a village in Rolpa where one last mass meeting has been organized. People set up an area outside with red banners on the walls and Party leaders, cadres, and peasants from the surrounding area start gathering in the late afternoon.

The local cultural team opens the program. As I watch a young dancer twirl gracefully to the music, I think about all the 'welcome' and 'farewell' programs that have been organized for me in the last several weeks and know that these meetings have not been easy to organize. People had to travel in dangerous conditions, sometimes walking for hours in the dark. And all this had taken place during the elections, a most intense period in which the government has flooded this region with thousands of police. Families have come to tell me about loved ones killed by the police – wanting the world to know about how the government is using brutal measures in its attempt to crush the insurgency. Squad members have come, fresh from encounters with the enemy, anxious to have me record their success in my notebook. Party leaders have spent hours with me, telling in great detail what it means to wage a Maoist People's War, hoping that this news will build understanding and support in other countries.

The program finishes at 8:30 pm and by this time everyone is ready to go inside to escape the darkness and cold and get a warm meal. Then a bunch of us sit around talking. The people I have been traveling with know that my time in the west is coming to an end and they have a long list of questions they want to ask me before I leave. After weeks of interviewing so many people, tonight it is my turn to answer their questions. They want to know what it's like in the United States. They want to know what life is like for people, what kind of struggle is going on, and if there is a chance for revolution in such a powerful country. They also want to talk about the international situation and the prospects for revolution around the world. Once again, it strikes me how much the revolutionaries here in Nepal see what they are doing as part of the worldwide struggle. In 1992, after the arrest of the leader of the Maoist People's War in Peru, Abimael Guzmán ('Chairman Gonzalo') in 1992, Maoists in Nepal organized mass demonstrations to demand his release. And I met many people who were hungry to hear news about the People's War in Peru. I also met several people who knew about the struggle to free Mumia Abu-Jamal, the political prisoner on death row in the United States.

The next morning there are several hours of interviews lined up for me. One man, from a leading area party committee in Rolpa, tells me a story about how people were able to develop the confidence to take up armed struggle:

> 'During Kilo Sera 2, 77 police came into this village and went from door to door looking for people. One area committee member and one Party member, two brothers, were arrested and tied up with rope. The younger brother got his hands untied and was able to run away. There were many police in the area and six police chased him. The older brother was shot and badly injured by the police, but he managed to escape and hid in a small cave. The police looked for him all day but couldn't find him. Party cadres knew about the cave, and at night they went to get him and took him to a safe place. The police blockaded the paths for two days and searched the village and the jungle but the comrade escaped. This incident made people see that it is better to resist when

the police come and not just let them arrest you. If the enemy comes viciously, we resist strongly. If we are afraid, the enemy will not retreat, but will step up their repression. During the government's Kilo Sera 2 campaign, the police came to attack us. But the masses resisted strongly, the enemy was forced to retreat, and now the place is in our hands.'

Twenty-nine-year-old squad commander Tamil sits down next to be interviewed and I ask him to tell me about his life, how he came to join the People's War. Over 90 percent of the people in Nepal are peasants. But the story Tamil tells me shows how some peasants have found their way into the revolution through their experience as workers:

'My family is very poor, so when I was 15, I went to India to find work. I lived there for eight years, working as a laborer in the countryside, collecting raw material for medicine. Later, I went to the city and worked in a factory making plastic bags. I also worked in a steel factory as a security guard and I worked in a chocolate factory and a pencil factory. The wages were very low, about 400 rupees (about $6.00) a month. Legally we were supposed to work only eight hours, but we had to work twelve hours. When we worked more than twelve hours we were paid very little for the extra hours.

'In 1994, I came back to Nepal and lived here one month. During this time I was in contact with the local Party leader, who told me about the Party. But at the time, I was not that influenced by them and decided to return to India. I lived in India this time for five months. I worked picking and hauling apples and then the season was over and I came back home in 1994.

'I visited the same Party leader when I returned, and we talked for a long time, many times, about the Party. The local leader talked about how the people are oppressed and how the usurers and other reactionaries in the village exploit the people. He also talked about how youth are exploited when they have to go to India to work. He talked about how we have to fight against the class enemy to make a better life. I thought about all this and became very influenced by the Party's politics and was interested to know more about the Party's plan to start a People's War.

'I got involved in the Party's activities and in 1995 I joined the YCL [the Young Communist League]. This was a time of sharp class struggle in this area, leading up to the Initiation, and I was very actively involved. The first action I was involved in was an attack on a local exploiter. I was charged with a case from this action and was forced to go underground.

'At the time of the Initiation, I was a member of a fighter group and was involved in the raid on the Holeri police post. Then in the Second Plan, I became a member of the first squad that was formed. I took part in the raid on the Piuthan Lung police post in 1997 and many other ambushes, mining actions, and arms seizures. Six months ago I became a squad commander.'

Several other people come by to talk and, by the time we are done, it is mid-afternoon and we are in a hurry to leave so we can travel in as much daylight as possible. We walk for about five and a half hours, and by the time we start looking for some place to stay for the night, it is very dark.

In Nepal there is a tradition among the peasants to give strangers food and shelter. Even if they are very poor, the peasants share what they have. The guerrillas sometimes will find a shelter where the people will welcome and gladly give food and a place to sleep to Maoists. But in this case, the people I am with don't want to reveal their politics since we are in somewhat unfamiliar territory, so they decide to say we are travelers who need some food and a place to sleep.

We knock on one door and two brothers let us into their small house. There is a fire going and we huddle round it to get warm. The meal is simple – wheat porridge and dried vegetables – but I'm very hungry and the meal tastes very good. After we eat, our hosts give us a few blankets to spread out on the floor and we lie down to sleep.

In the morning we thank our hosts and give them some rupees. Then we are back on the trail again by 5:15 am and reach our destination by 9:30 am. When we first arrive, Rachana, the squad member who was the first woman guerrilla to be my aide, comes out to greet us. I am very surprised to see her. Then inside, there are more reunions. The Central Committee member in charge of this guerrilla zone, the District Committee secretaries of Rolpa and Rukum, and others whom I have met with over the last month have gathered here for a final evaluation meeting.

Some other Party leaders have also come to be interviewed, including the District Committee secretary of Jajarkot, one of the districts here in the west where the People's War is very strong. I didn't have time to visit Jajarkot, which is northeast of Rukum, so the DCS has traveled many hours to come tell me about the work in his district. He says:

> 'The main political contradiction in our district is between the people and the ruling party, the Nepali Congress [NC]. And with the revisionist UML [Communist Party of Nepal, Unified Marxist-Leninist] becoming closer to the NC, things are also sharpening up with them. But supporters of the NC are becoming attracted to the CPN (Maoist), and there are inner-party contradictions within the NC, UML and RPP [the pro-monarchist Rashtriya Prajatantra Party].
>
> 'There has been vicious oppression in Jajarkot. The government claims 50 percent voted in the elections but in reality much of the votes were forced by the police, people were afraid.
>
> 'The main economic contradiction in our district is between big usurers/cheaters and the people. There are hardly any landlords in our district and because of the People's War, many of the local reactionaries have left the villages or become inactive. But the government still uses these forces against the people. They live in district headquarters but sometimes return to cause trouble for the people. The government is also using new forces – police come into the village and try to force people to act as spies. The Party has been able to stop some of these people through struggle or with threats. We take action against those who don't stop working for the enemy.'

The DCS tells me that in Jajarkot, as in Sallyan, the influence of the Party was not very strong before 1990. He says a lot of good work was done after 1990 and the Party and mass organizations grew, leading up to 1996, when the People's War started. After the government responded with vicious repression, many people became afraid and stopped supporting the Party. But, he says, they were able to reverse this, and one of the key things in turning the situation around was organizing the people to hit back at the enemy in a well-planned, effective, and strong way. He describes some of how this dialectic between revolutionary moves and counter-revolutionary moves has gone:

> 'In March 1998, there was an ambush by the People's Army in district headquarters. One policeman was killed and three others seriously injured. During Kilo Sera 2, in Ragda VDC the People's Army did a mining ambush in which two policemen were killed and two seriously injured. During an election, a squad attacked a NC propaganda team and killed

one and injured two. In an ambush in Laha VDC, one policeman was killed and one was injured. There have been many other smaller actions by the People's Army. Since the initiation 15 reactionaries, mainly spies, have been annihilated.

'There is feudalistic thinking among some people, and some people are related to or have ties to the Royal Palace and have influence on the masses. There is also some religious belief in spirits and gods and witches. This kind of thinking is stronger in Jajarkot than in Rukum and Rolpa, and this creates some political difficulties. Most people have some religious belief, even if they don't support the government. People accept the economic concept of communism more easily than the ideology of communism. And the reactionaries wage propaganda against Maoists, saying, "They don't believe in God, they destroy temples, etc." Among the masses there is some resistance to breaking feudal traditions, but gradually people are accepting new things.

'The method of attack by the police is to encircle and surprise. On one day police posts combine to attack a particular village and then on another they will do the same to another village. The government sees Jajarkot as a doorway to the far western part of the country. This area is geographically favorable to the People's War and the police are burning the forests here as a way to attack the revolution.

'The government repression has affected our ability to get shelter as well as our communications and this is a big problem. Some people who have been afraid of police repression have left and gone to India. If the police know about people who give Maoists shelter and food, they will arrest and/or kill them. The families of full-timers are constantly harassed by the police. The police capture food grains from the farms of full-timers. Police also confiscate the land of families and tell the people they will be killed if they farm the land. Then the family has to leave or become more involved with the Party. Many times, the young people join the People's War while the old people and children leave and live somewhere else. The Party is helping some of these people. But it is hard to solve this problem because there are more and more people like this in need of support. If we wage strong military actions and create a power vacuum, this problem will be solved with people's power in our hands.'

* * *

Today is an easy day, a respite from the intense conditions and pace of our journey. In the afternoon, as the sun dries our clothes, we all have a chance to relax. Many of the guerrillas are sitting around and reading. Someone has also brought a small collection of Xeroxes that people have been studying. There are readings from bourgeois military books, mainly from India, but also one from the US. There are several books on the history of the Chinese Revolution, including William Hinton's *Fanshen*. And there are also two books by Maoists in the United States – *And Mao Makes 5* (a collection of documents and articles from the Cultural Revolution in China, edited by Raymond Lotta) and *Phony Communism is Dead ... Long Live Real Communism* by the Chairman of the Revolutionary Communist Party, USA, Bob Avakian.

Police in the guerrilla zones don't allow revolutionary newspapers into the area. You could be arrested, possibly shot, for bringing any kind of Maoist literature into Rolpa. But someone had smuggled in an issue of *Janadesh*, one of the Maoist newspapers printed in Kathmandu. When they pull it out of their pocket, a cheer goes up. The issue is a couple of weeks old, but the guerrillas eagerly pass it round, poring over every article, reading some parts aloud. Battle news from around the country and revolutionary analysis and commentary provide a lifeline for these rebels. All afternoon, the already tattered newspaper, limp from being folded and refolded so many times, is passed round the room and is then carefully put away so it can travel to another village.

My time here in the west, the eye of this storm of the People's War, has been quite an experience in so many ways. I feel like I have not only learned a lot of basic facts about how the People's War is advancing, but I have also been able to get a real sense of the human side of this revolution – by traveling, living, and talking with members of the Party, the People's Army, and peasant villagers. And being in the west has given me a taste of war conditions.

Most people around the world don't know about the war going on here. Nonetheless, what I have seen here seems to be extremely significant and relevant to the situation of hundreds of millions of people. At the end of my interview with Prachanda I ask him how he sees the importance of the People's War in Nepal in relation to the rest of the world and he says:

> 'Nepal is a small country; we are a small party. But we have a big perspective. Our People's War may be a spark, a spark for a prairie fire ... This war has changed the name of the country itself – the identity of the country. It was a very backward, poor, and beggar country. But

now it is a country of heroes, of proletarian heroes. And now on the world's highest peak, Sargamatha [Mount Everest], the red flag is there. This will be seen from all over the world. People will say: What country is Nepal? It is the country with the world's highest peak, Mount Everest. What is there? Heroic proletarian revolution, People's War is there. This will be seen ...

'When you are in a pond or in the middle of a lake you do not know the importance of water. But when you are in the desert, then you see that just one glass of water is very important. Today there are not many genuine People's Wars in the world. So in this desert of revolutionary war, the People's War in Nepal is one glass of water for all the revolutionary people. And we will fulfill our duty to give water to the revolutionary people.'

* * *

In our evaluation meeting, people ask me what I have gotten out of this trip. I tell them how I have learned a lot about the role of the youth, women, and the oppressed nationalities. I talk about how the families of those killed by the police seem determined to win, even in the face of vicious repression. And I tell them that I see that they are trying to wage a revolution that is not only about destruction of the present society, but also about what they see as revolutionary *construction* of a new way of life. I have seen how they are trying to develop the seeds of a new society, how people are breaking with feudal traditions and social relations and creating a new economy and culture.

I also tell people that I have learned a lot about how they have advanced the military theory and practice of waging their People's War – in just a short period of only three years. They have gone from primitive fighter groups to squads, platoons, and larger task forces. They have developed from small attacks to more developed military raids and ambushes.

We talk about how the situation is at a very critical point right now. The government is compelled to step up their efforts to crush the People's War, exactly because the revolution has continued to grow. The government is especially worried about the development of power vacuums in areas where government officials have fled and the police are afraid to come in. The police posts are being centralized in many areas, which will mean that the guerrillas will be confronted with larger groups of police. So far, the national police have been the ones sent against the People's War, but there has been talk in the government about sending in the Royal Nepal Army.[1] All this will require a leap in the military capacity of the People's Army – in the size of military groups (from platoons to companies and even larger); in the level of military actions; and in the quantity and quality of weapons. I think that in the near future there will be some major moves by the government against the People's War and this would demand a big leap in the capacity of the People's Army as well as the whole party leadership, mass organizations, etc. This would also make it more urgent for there to be international awareness of the situation.

Before we leave this last village, people gather outside for a final farewell. People remind me that I am the first foreign journalist to be allowed this kind of access in the guerrilla zones, and they reiterate how important it is now for me to make their story known to the world.

We leave at noon and, a bit later, arrive at the village school where about 100 people have gathered. The local villagers heard that we were going to come through here on our way out and they want to greet me. The local party leader presents me with some farewell words and gifts. Then we shake hands with everyone and many step forward with garlands of flowers to put around our necks.

On our way out of the Rolpa/Sallyan districts, the people I am with point to a tall, pointed mountain peak, not too far away. It is where the government's Jhimpe Communications Tower used to be before the People's Army raided it, injuring two police and seizing a number of rifles and ammunition. As the sun begins its descent over this scene, this seems a fitting image for my exit and I mark the sight, sound and feel of this moment in my memory.

As we leave Rolpa, I turn to look back up at the towering countryside and I think about all the dreams of the peasants, deep in the crevices of this incredible terrain. Geologists have reported that every year the Himalayan range pushes further up towards the sky. Now, in the foothills of these majestic mountains at the top of the world, another force is radically changing the landscape.

Afterword

When I was in Nepal in 1999, the Communist Party of Nepal (Maoist) had been leading an agrarian-based armed struggle against the Nepalese government for only three years. The guerrilla forces were still mostly armed with primitive weapons and an occasional modern rifle seized from the police. They were carrying out relatively small actions – raiding isolated police posts and attacking hated landlords and corrupt politicians. Such actions sometimes led to the seizure and redistribution of land and were popular among many impoverished peasants who saw their oppressors run out of the villages and who benefited from the social, economic, and political transformations made in the areas under Maoist control.

The fighting units at that time were squads of seven to nine guerrillas and platoons with 24 to 30 fighters. In the Maoist stronghold of Rukum, I visited a camp of 'special forces' which had been formed to carry out larger actions during the boycott of the elections. This was part of an attempt at the time to form company-size units of about 100 guerrillas. By September 2001, when the People's Liberation Army (PLA), was formally declared, it had several permanent companies, and in some cases was fighting in units of brigade strength – several hundred soldiers. Today, after eight years of fighting, the People's Liberation Army has many modern weapons, including GPMGs, LMGs, SMGs, SLR, and rocket launchers.

At the time of my trip, the Maoists were also in the beginning stages of establishing political authority and organization in the villages. The party had divided the areas where they were active into three zones: (1) 'Guerrilla Zones,' where there was a police presence and fighting and armed resistance was taking place; (2) 'Propaganda Zones' in urban areas, where the regime's power remains strong and the main aim is to prepare the ground for eventual insurrection – and the main forms of work are political education, mass activities, and building support for the struggle in the countryside, including among the middle classes; and (3) 'Main Zones,' where they had not yet, but were in the process of establishing 'base areas' that would serve as embryos of 'red political power.'

Less than three years later, by the end of 2002, the People's War had made impressive advances, expanding the areas under its control, as well as developing new institutions of 'people's power.' The government structure of VDCs (Village Development Committees) had broken down in much of the countryside. Elected VDC chairmen had either left the area or quit their posts and were working with the Maoists. Politically, the government had little if any presence and authority in the countryside. The Maoists had established base areas not just in the Western Region, the stronghold of their revolution, but also in the Eastern and Central Regions. They stated that in the Western Region, ten million people – out of Nepal's total population of 23 million – lived in areas under their control. Reports in mainstream newspapers and intelligence reports admitted that the Maoists were active in most of Nepal's 75 districts.[1]

GUERRILLA STRATEGY

How did the Maoists make such impressive military and political gains in only three years?

In developing the fighting capacity of its guerrillas, the Communist Party of Nepal (Maoist) applied the principles of Mao Tsetung's military strategy – tactically pitting 'ten against one' and strategically 'one against ten.'[2] For example, they intensified their military assaults on weak links of the government, mainly the less fortified police posts.[3] They recognized that on a nationwide level the revolutionary forces were (and would remain) outnumbered by government forces, and so, in an overall strategic sense, faced 'one against ten.' But tactically, and in particular battles, they saw it was possible to concentrate an overwhelming superior force to wipe out concentrations of government forces – with an orientation of 'ten against one.' To achieve this, larger and larger units of PLA forces were developed, first by bringing together three platoons (about 30 guerrillas) to make companies (of 100 guerrillas) and later three companies to make brigades. At the end of November 2001, when the Maoists terminated a four-month ceasefire, they launched an action that involved one battalion, two additional companies, and some platoons – a total of 1,335 from the PLA and 700 militia.

The CPN (Maoist) consciously applied Mao's strategy of protracted warfare. For example, when the government tried to provoke the Maoists into a situation of a 'fight to the finish,' the PLA avoided all-out battles and instead took the approach of waging guerrilla warfare, luring the government forces deep into 'red areas,' encircling them and striking big blows at their weakest links. In these red areas, broad popular support has provided the Maoists with intelligence and reconnaissance, and local militias play an important political and military role. As military units of the People's Army have grown and the level of warfare advanced, smaller units and guerrilla actions have continued to play an important role.[4]

In this way, the guerrillas were able to successfully carry out military actions, even with primitive weapons and relatively small fighting units, and the police were increasingly put on the defensive and eventually forced to stay holed up in their barracks most of the time. By the end of 2002, government forces didn't control much outside of the district headquarters and some parts of the Terai, along the southern border with India.

The concept of establishing base areas and new political power has been part of the program of the CPN (Maoist) from the very beginning of the war and is seen by them as a crucial component of their overall strategy.

As the police, officials, and landlords were driven out of the countryside, a new situation developed in which the authority and institutions that had ruled over and oppressed the people ceased to exist. The Maoists stepped into this vacuum and set up a 'new people's power.' In 1999, the guerrillas were just beginning to do this, organizing local committees to run village life and 'people's courts' to settle disputes, distribute land, grant divorces, punish rapists, etc. These were the beginning steps toward establishing functional base areas.

The ability to consolidate such initial political advances into real base areas was directly related to the military advances being made. That is, the more the military struggle was able to 'liberate' territory and actually carve out areas where the police and other government forces dared not enter, the more the Maoists were able to consolidate their political authority in a more ongoing, even if still relatively tenuous, way.

In all of the guerrilla zones I visited in Rolpa and Rukum, the Maoists had clearly established support and were able to walk round in the villages in full uniform. But travel through much of these areas at that time still required that the guerrillas wear civilian clothes and travel in the dark of night. And it was still the case that the police were able to come in and out of these areas, making it very dangerous for the guerrillas to stay in a village for more than one or two nights at a time.

In early 1999, the CPN (Maoist) was still in the process of carrying out its 'Fourth Plan' of beginning to establish base areas. Two years later, they had strong base areas in the western districts and were accelerating the expansion and consolidation of base areas in other parts of the country. In the districts of Rukum, Rolpa, Jajarkot, and Sallyan in the Western Region of the country, 'United Revolutionary People's Committees' were openly exercising power and functioning as the main organs of administration. 'People's committees' were acting as the embryo of new local governments, running administrative, economic, social, cultural, education, and development departments and responsible for local militias, 'people's courts' and 'people's jails.'

All these military and political advances by the Maoists were quite significant. But after five years of fighting, by the beginning of 2001, they had not yet directly confronted the main forces of the government – the Royal Nepal Army.

REVOLUTION AND COUNTERINSURGENCY

As it became clear that the police forces could not stop the guerrillas, sections of the Nepalese ruling class and foreign powers, including the US and India, exerted pressure on the king to mobilize the army against the Maoists.

Under Nepal's Constitution the king effectively controls the Royal Nepal Army (RNA). But King Birendra was reluctant to unleash the RNA against the guerrillas, which only intensified the infighting within Nepal's ruling class – both between the palace and the parliamentary forces, and within the parliament itself.

This was the situation leading up to the shocking events of June 1, 2001, when the king, queen, and most of the royal family were massacred in the Royal Palace.

Few believed the official story that the king's son went on a shooting spree because his mother didn't like his girlfriend. The king's brother, Gyanendra – who would prove to be much more willing to mobilize the RNA – was conveniently on vacation the night of the murders and took the throne under a cloud of suspicion.

Prime Minister Girija Koirala was then forced to resign after a guerrilla action in which a number of police were taken prisoner and RNA units nearby failed to come to their rescue – highlighting the fact that the army was being restrained. Gyanendra chose a new prime minister, Sher Bahadur Deuba, who called for negotiations with the Maoists. In July the CPN (Maoist) agreed to a ceasefire and several rounds of talks, in which they presented their demands for the abolition of the monarchy and the establishment of a republic; the formation of an interim government; and a popularly elected constituent assembly to draw up a new constitution.

During this ceasefire, both sides carried out intense preparations for the resumption of fighting. The government stepped up efforts to acquire modern weapons and foreign aid, while the Maoists used the break in fighting to mobilize wider support and train its forces to carry out more sophisticated military actions. Toward the end of 2001, the Maoists announced the formation of the 'People's Liberation Army' and the United People's Revolutionary Council, which would now govern areas under Maoist control.

On November 21, 2001, Prachanda, the head of the CPN (Maoist), issued a statement saying there was no reason for further talks since the government had rejected the main demands of the party. Two days later the ceasefire came to a dramatic end when the guerrillas carried out actions in more than 20 of the country's 75 district headquarters.

King Gyanendra declared a State of Emergency and for the first time fully unleashed the Royal Nepal Army against the guerrillas. He suspended many constitutional rights, ushering in months of repression with widespread arrests and, in the countryside, the killing of thousands of suspected 'Maoist sympathizers.' Censorship of the press was harsh and widespread. Newspapers openly sympathetic to the Maoists were forced to close, their editors and journalists arrested. The Maoist newspaper *Janadesh* was raided and shut down and its editor Krishna Sen was jailed and later murdered in custody. The government also went after mainstream newspapers that wrote anything about the Maoists the government didn't like. Editors and writers were interrogated simply for printing quotes from Maoist leaders. In August 2002, the Federation of Nepalese Journalists reported that over 130 journalists had been arrested since the State of Emergency was declared.[5]

The resumption of fighting between the PLA and the government came less than two months after the September 11 attack on the World Trade Center in New York. America's 'war against terrorism' created a convenient rationale for the Nepalese regime to enlist greater international support for its fight against the guerrillas. The Nepalese government joined India in putting the 'terrorist' label on the CPN (Maoist) and went on a concerted campaign to get support from other countries. In January 2002, US Secretary of State Colin Powell went to Nepal and met with the king, prime minister, and top military generals. Commenting on this visit, the *Kathmandu Post* got at something real when it said, 'Call it the "September 11 syndrome," but the fallout from America's global war against terrorism has helped to line up important international constituents behind Mr. Deuba's [Nepal's prime minister] own war against terrorism.'[6]

The US and the UK led the way in stepping up political and military support for the embattled Gyanendra government. Soon after Powell's trip the British Chief of Defence Staff, General Michael Boyee, also visited Nepal to meet with army personnel and the king, and toured RNA bases in the Western and Eastern Regions.[7] In May, Prime Minister Deuba went to the US to meet with President Bush and then flew to London to ask for help there too.

Britain provided $40 million and the US came up with $22 million in aid. The US also sent a dozen military experts who surveyed different parts of Nepal in order to map out operational plans for the RNA. In June 2002, the British government hosted an international meeting to discuss how to help the Nepalese regime. Mike O'Brien, the Under-Secretary of State for the Foreign and Commonwealth Office left no doubt about the position of the British government when he said, 'It is imperative that we help the government of Nepal in its struggle against terrorism. We cannot allow the terrorists to win. Nepal must not be allowed to become a failed state.'[8] In October, O'Brien convened a follow-up meeting in Kathmandu.

India also moved to shore up the Nepalese regime by providing truckloads of military hardware and helicopters, moving its own troops closer to the border, and clamping down on Nepalese living in India.[9] Indian authorities arrested Nepalese journalists and 'suspected Maoists' and extradited them to Nepal where they were held without trial. They arrested wounded Maoists from Nepal who were being treated at private hospitals in India and handed them over to the Nepalese police.[10]

In 2003 the Indian government stepped up efforts to hunt down and arrest leaders of the CPN (Maoist). On August 20, 2003, Chandra Prakash Gajurel, known as Comrade Gaurav, a member of the Political Bureau of the CPN (Maoist), was arrested by the Indian authorities at Chennai (Madras) airport as he attempted to travel to Europe to carry out Party work. And on March 30, 2004, Mohan Baidya, a senior leader of the CPN (Maoist), was in Siliguri, in West Bengal, undergoing treatment for cataracts, when he was arrested by Indian police. Baidya was charged with conspiracy and firearms violations, and his assistant, Narayan Bikram Pradham, reported to be an Indian of Nepali origin, was also seized by the Indian authorities.

The Nepali government offered cash rewards of NRs 1,000,000–5,000,000 (US$13,000–65,000) to anyone turning in top Maoist leaders, dead or alive.[11] And in response to a request from the Nepalese government, the International Criminal Police Organization (Interpol) issued 'Red Corner Notices' (RCN) against top Maoist leaders, authorizing arrest warrants in all 179 Interpol member countries.

The State of Emergency was formally lifted after seven months, but widespread arrests and killings by government forces continued. In the year after the State of Emergency was declared, the police and RNA killed more than 5,000 people.[12]

Then, with the ruling class still fractured over how to deal with this escalating situation, the king made a drastic move. On October 4, 2002, Gyanendra seized all executive power, dismissed the Deuba government, and indefinitely postponed the national elections originally scheduled for November.

This extreme fissure in the ruling class and the complete breakdown of the parliamentary structure was the situation leading up to the second round of negotiations with the Maoists, which began at the end of January 2003.

The CPN (Maoist) knew that it could enter into negotiations from a position of strength due to the military and political advances it had made. On the other hand, the government recognized they could not win through military means alone and hoped that negotiations might provide some maneuvering room to disarm and destroy the Maoists.[13]

Several major leaders of the Party who had been underground for years emerged to negotiate with the government. At the same time, the Maoists utilized the ceasefire to hold huge public rallies in Kathmandu and throughout the country. Top Maoist leaders like Baburam Bhattarai set up an office in Kathmandu, made themselves available to the press, and gave public speeches urging people to support the revolution's demands.

King Gyanendra continued to exercise absolute control, appointing a prime minister from the pro-monarchist RPP (Rashtriya Prajatantra Party). Meanwhile, the Nepali Congress, which had been the ruling party, and all the other parliamentary parties, were completely cut out of the government and the negotiations. The five main political parties called for a boycott of the government and held a series of street demonstrations against the king. Hundreds were arrested, including senior leaders of the Nepali Congress and the parliamentary-based Communist Party of Nepal (United Marxist Leninist).

This made for a very complex situation, as both parties had been part of the government before the king dissolved parliament and in various ways had been directly involved in the counterinsurgency campaign during the 2001–2 State of Emergency. This sharp division within the ruling class helped the Maoists, who took the opportunity of the ceasefire to meet separately with leaders of the parliamentary parties.

Throughout this round of negotiations the US provided military aid and training to the RNA, delivering 5,000 M-16 rifles with the promise of 8,000 more. Development grants for projects termed 'insurgency-relevant' were increased from $24 million to $38 million. The US also pushed through a five-year 'anti-terrorist' agreement with Nepal in which the US will provide arms and training to counterinsurgency forces. As if to make a point, even after the Nepalese government withdrew the 'terrorist' label it had placed on the Maoists as part of the conditions of the ceasefire, the US added the CPN (Maoist) to its State Department 'terrorist watch list.'

In six months, there were two sessions of talks between the Maoists and negotiators for the king. Then, on August 17, 2003, as a third meeting was taking place, RNA soldiers murdered 19 Maoists in a village in the eastern district of Ramechhap. According to Amnesty International, security forces opened fire on a house where people were meeting, one Maoist was killed and the other 18 were taken away and later lined up and executed.[14]

Soon after this, Party leader Prachanda issued a statement denouncing the 'cold-blooded killings' and the government's refusal to seriously discuss the Maoists' main demands. (Negotiators for the king had declared they would never accept the end of the monarchy through a constituent assembly and the establishment of a republic, and demanded that the PLA hand over their guns.) The next day, August 28, the Maoists ended the ceasefire with several military actions across the country.

Within days the US and UK moved to intervene more directly. The American and British ambassadors went to the home of Nepali Congress leader Girija Koirala and asked him to unite with the monarchy and the government to fight the Maoists, arguing that opposition to the king would only strengthen the Maoists.[15] The

US and UK ambassadors also visited the head of the CPN(UML) to deliver a similar message.

As the People's War has continued to advance, the Nepalese ruling class has been fraught with deep and seemingly irreconcilable divisions. The deep economic and social inequalities of Nepalese society continue to fuel the Maoist revolution. The parliamentary parties are widely seen as corrupt and ineffective. The king's move in October 2002 to dismiss parliament and usurp power eliminated any pretense of a democratic system – leading to even deeper splits within Nepal's ruling class and further cynicism among the people. The RNA has been bolstered with thousands of additional soldiers, training from the US, new modern weapons, and helicopters, but has still been unable to prevent the Maoists from gaining control of most of the countryside. At the beginning of 2004, the Maoists announced they had control of 80 percent of the rural areas.

All this has caused increasing concern – and led to increasing intervention – by the US, the UK, India, and other powerful countries.

POST-SEPTEMBER 11 AND THE GEOPOLITICAL STAKES

Colin Powell's visit to Kathmandu came four months after September 11 and two months after King Gyanendra declared a State of Emergency. This was the first high-level diplomatic trip by a US official to Nepal in 30 years and signaled mounting concern, necessity, and willingness by the US to provide not only political but also military support to crush the People's War.

The US doesn't have lots of investments or sweatshops in Nepal and there aren't any significant oilfields in this small Himalayan country. So what's behind this high-level concern? Why is the US providing the Nepalese army with millions of dollars, thousands of weapons, and military advisers and trainers? Why has the US, as a July 1, 2003 article in the *Kathmandu Post* revealed, been 'quietly securing close military and political ties with Nepal'?

Since September 11, the US 'war against terrorism' and the aims and ambitions of the US crusade to attain unrivaled world hegemony have been setting the terms for much of international relations, including how the US (and other powers) look at their necessity and freedom to intervene in Nepal.

As part of the US quest for world domination, the 'war on terrorism' serves as an all-purpose umbrella under which numerous interventions are being justified. The political and ideological program of the

Maoists in Nepal clearly has nothing in common with the reactionary politics and religious fundamentalism of groups like al-Qaeda. But this hasn't stopped the US from using the pretext of 'combating terror' to justify military action against any and all insurgencies which threaten US interests – including genuine revolutions aimed at overthrowing oppressive governments.

The US, Britain, and other imperialist powers have provided the Nepalese regime with political and military support exactly because they know that a Maoist victory would reverberate throughout the Indian subcontinent and the world. This is a region of extreme instability where a Maoist 'regime change' in Nepal could interact in unpredictable ways with the hostility between Pakistan and India, the conflict in Kashmir, relations between India and China, and other guerrilla insurgencies in the region, especially those in India.

The often antagonistic relationship between India and China is certainly a factor in this developing scenario. Nepal is strategically situated between the Tibetan region of China and the northern border of India. Because of this, both of these major powers view Nepal as a kind of 'buffer,' over which each has jockeyed for influence and power as a way of challenging and defending against the other.

India would be seriously threatened by a government in Kathmandu run by Maoists (who have already stated that one of the key goals of their revolution is to end Indian domination). And the New Delhi government worries that China would try to take advantage of any kind of upheaval in Nepal to strengthen its hand against India and in the whole region.

The Maoists in Nepal denounce the current Chinese regime for overthrowing and dismantling socialism after Mao's death. The Chinese government has made clear that it supports the efforts to crush the insurgency, but China would be extremely concerned if India invaded Nepal to prevent the Maoists from seizing power. This would upset the long-standing and fragile balance of power on the China/India border, where serious warfare has broken out before and possibly provoke renewed hostility between these two major powers.

In terms of the US and its geostrategic stakes in Nepal, the question is: Can the US in its quest to achieve unprecedented global hegemony allow a successful Maoist guerrilla war in Nepal?

In a radio interview at the end of 2003, Michael Malinowski, the US ambassador to Nepal, revealed something about how the US views this question.[16] He said:

> 'It's a troubled country. It's very disturbing. We're concerned about it. One may ask why does the United States care, it's 8,000 miles away? I would say there's a number of reasons. One on the ideological plane we want democracy to succeed. We don't want to see democracy fail. We don't want to see democracy fail by a group, a small group that is unwilling to contest its ideas in the electoral process or the parliamentary process. But instead have decided to go the way of the gun, use terrorism, terrorist acts to get their will ... There's real reasons why people have picked up the gun here. They're impoverished. There's a lack of access to higher levers of education. There's corruption. There's mismanagement. There's bad government. All of that. But again I would argue that Maoism is not the way to solve that.'

Later Malinowski spoke to the international dimension of the US concerns, saying that he sees the need for 'A clear message from every outside nation or people who care about Nepal – a clear message to the Maoists that the world will not put up with this type of behavior and indiscriminate use of terror.'

Malinowski's statements deliver both a message and threat: this Maoist revolution is 'not the solution' and 'will not be tolerated.'

While Malinowski tars the Maoists with the 'terrorist' brush, he can't offer any real evidence to justify the charge. In fact, he is forced to admit that the Maoists have real support because of the oppressive conditions in Nepal – and that this is the basis for the military and political strength of the Maoists.

On one level, it seems almost counterintuitive that an operative of the Bush administration would be conceding that Maoist guerrillas have a base among the people and that poverty, deep inequalities, and a corrupt and bad government are fueling the revolution. While Malinowski calls the Maoists 'terrorists' he cannot simply write them off as 'terrorists' and has to acknowledge that they are actually seen by *millions* of people as a real alternative to the current order. Malinowski's arguments tell you something about the nature of this insurgency – that this is not an uprising of isolated rebels, but a revolution with mass, popular support that propounds a viable program for how to run Nepal and is contending for power.

The US fears that a Maoist victory in Nepal would, to use a phrase from Mao, 'ignite a prairie fire.' Intelligence reports assessing the impact of the insurgency in Nepal point to growing Maoist guerrilla warfare in India.[17] And the Naxalbari area, where Maoist armed struggle began in India in the 1960s, is located right across Nepal's eastern border.

In June 2002, at the request of nine Indian states where Maoists are waging armed struggle – Andhra Pradesh, Bihar, Chattisgarh, Jharkhand, Madhya Pradesh, Maharastra, Orissa, Uttar Pradesh, and West Bengal – India created a special police force to crack down on Maoist guerrillas.[18] Official Indian sources have said that any negotiated deal in Nepal that produces a government with 'unreformed' revolutionary Maoists sharing power would have serious security consequences for India and its war against Indian Maoists.[19]

CAN THE REVOLUTION WIN?

The political and geographic connections and interrelationships between the situation in Nepal and India (and the whole region) also have huge implications for and present real challenges to the Maoists if they do seize power in Nepal.

Militarily, a new revolutionary government would immediately face the problem of being surrounded by unfriendly governments and the possibility of military invasion – by India, UN 'peacekeeping troops' or even the United States.

Economically, as an underdeveloped country with a long history of dependency on India and other foreign countries, and an almost complete lack of industry, the new Nepal would immediately face enormous challenges in meeting people's basic needs and developing an economy that does not create relations of foreign dependency and exploitation.

Politically, the Maoists would be attempting to build a new socialist society surrounded by hostile states, in a world where 'communism has been declared dead,' and in which there are tremendous prejudices against socialist states led by communist parties. During the Chinese revolution Mao had to deal with what he called 'bourgeois democrats becoming capitalist roaders' – that is, those within the Communist Party itself who united with the anti-feudal, democratic aims of the revolution but then became proponents of building a *capitalist*, not socialist, society. This would certainly be a phenomenon in Nepal, the outlines of which can already been seen in the outlook of various political parties that are proponents of bourgeois democracy, not socialism, but who could unite with the Maoists to oppose the monarchy.

Bob Avakian sheds valuable light on this complex relationship between seizing power in one country and promoting the advance of revolutions elsewhere. He says:

> 'Socialist countries have so far emerged and for a certain historical period are very likely to emerge one or a few at a time. So, in grand strategic terms socialist countries and, more broadly, the international proletariat and the international communist movement will be faced with a situation where it is necessary to change the world alignment of forces or face the prospect of socialist countries going under after a certain point. This doesn't mean there is some sort of mechanical mathematical or arithmetic equation where if you don't get more and more of the world in a given period of time, then the socialist country, or countries, that exist at the time (if there are any socialist counties right then) will inevitably go under. But there is a contradiction when a socialist country is in a situation of being encircled; and that also interacts with the internal contradictions within the socialist society. And, at a certain point, if further advances aren't made in the proletarian revolution worldwide, these things will turn to their opposites and the conditions will become more favorable for capitalist restoration within the socialist country. This doesn't mean capitalist restoration automatically kicks in after a certain point, or that it will automatically occur at all. But it means that things will begin to turn into their opposites and the conditions for capitalist restoration will become more favorable. So, in that dialectical materialist sense, it's one way or the other: make further advances and breakthroughs in the world revolution or be thrown back, temporarily.'[20]

Recognizing the greater regional and world significance of the People's War in Nepal also informs the overall strategy of the CPN (Maoist) – in how it sees the path to seizing power *and* in how it sees consolidating state power. The Party leaders very consciously look at their revolution as 'part of the world revolution' and they look at the success of a revolution in Nepal as providing both an example and a 'base area' for further Maoist revolutions.[21] This has huge implications for the armed struggles being waged by Maoists in India and in turn reacts back on the struggle in Nepal. When I interviewed Prachanda he emphasized the importance of the struggle in India to the success or failure of the Nepalese revolution – pointing to the positive factor of a reinforcing synergy between Nepalese Maoists living in India, the political forces in India that would oppose intervention in Nepal, and the overall progress of Maoist revolution in India.[22]

The CPN (Maoist) emphasizes its 'proletarian internationalism' and has given this organizational expression through its participation in the Revolutionary Internationalist Movement (RIM), whose participants include Maoist parties and organizations from around the world. The CPN (Maoist) also makes no secret of the fact that it has fraternal relations with Maoist organizations throughout the Indian subcontinent. For example, in June 2001, the CPN (Maoist) helped form CCOMPOSA (Coordination Committee of Maoist Parties and Organizations of South Asia), which is made up of ten parties, including ones from Bangladesh, Sri Lanka, and India. CCOMPOSA's stated purpose is to 'unify and coordinate the activities of the Maoist parties and organizations in South Asia to spread protracted people's war in the region.' Maoist parties in India have clearly been encouraged and emboldened by the success of the Maoists in Nepal. A press statement announcing the formation of CCOMPOSA emphasized how the struggle in Nepal is 'changing the political geography and revolutionary dynamics of South Asia.'[23]

All this underscores the strategic significance and political importance of the Nepalese revolution – and the fact that the US and other imperialist powers cannot (and will not) ignore the real threat of a Maoist victory in Nepal. This also poses critical questions if the revolution does come to power in Nepal – what this could spark in neighboring countries, how the US, UK, India, and other countries would respond, and what kind of support would have to be built in the world for a new Maoist government in Nepal to survive the pressures and attacks of surrounding states and America's new imperial order. This would be a real challenge, not just for the revolutionaries in Nepal but for all who stand against injustice and oppression.

Notes

INTRODUCTION

1. *Washington Post Foreign Service*, June 2, 2001, 'Prince Kills 8 in Nepal's Royal Family Suicide Rampage Linked to Dispute over a Bride,' by Pamela Constable.
2. Li Onesto, 'Nepal: Intrigue and Insurgency,' *San Francisco Chronicle*, June 13, 2001.
3. Baburam Bhattarai, 'Political-Economic Rationale of People's War in Nepal,' *The Worker* #4, Organ of the CPN (Maoist), May 1998; the Third UN Conference on the Least Developed Countries, Nepal Country paper, Brussels, May 14–20, 2001.
4. Martin Hoftun, William Raeper, and John Whelpton, eds., *People, Politics and Ideology, Democracy and Social Change in Nepal* (Kathmandu: Mandala Book Point, 1999), pp. 72–4, 267–73.
5. Ibid., p. 263.
6. Bhattarai, 'Political-Economic Rationale of People's War in Nepal'; 'Revolution in Nepal, a Better World's in Birth,' *A World To Win* #29, 2002.
7. 'Revolution in Nepal, A Better World's in Birth.'
8. Li Onesto, 'Red Flag Flying on the Roof of the World' (interview with Comrade Prachanda), *Revolutionary Worker*, February 20, 2000.
9. The *Declaration of the Revolutionary Internationalist Movement* was adopted in 1984 by the founding conference of the Revolutionary Internationalist Movement. It is available on the web at awtw.org/rim
10. *Strategy and Tactics of Armed Struggle in Nepal*, document adopted by the Third Plenum of the CC of CPN (Maoist), March 1995, *The Worker* #3.
11. See Mao Tsetung, *Selected Military Writings* (Peking: Foreign Languages Press, 1968) and Bob Avakian, *Mao Tsetung's Immortal Contributions* (Chicago: RCP Publications, 1979).
12. Onesto, 'Red Flag Flying on the Roof of the World.'
13. *FY 2002 Foreign Operations Emergency Supplemental Funding Justifications* available at: http://www.fas.org/asmp/profiles/aidindex.htm
14. *South Asia Intelligence Review*, Volume 2, No. 21, December 8, 2003, points out: 'The withdrawal of the agencies of the state is complemented by the widening sphere of Maoist presence and activity. Towards the latter half of 1999, 20 of Nepal's districts were considered "seriously affected" by Maoist violence. By 2001, their number was up to 68. Now, all 75 districts in the country, including the capital, Kathmandu, have witnessed significant Maoist violence.'
15. Data taken from United Nations Development Programme, *Human Development Report, 1999*; Food and Agricultural Organization, *State of Food Insecurity in the World, 2003*; World Health Organization, 'Urgent Call to Improve Survival of Millions of Children' (June 27, 2003).

2 VILLAGES OF RESISTANCE

1 See photo 3.

4 RIFLES AND A VISION

1 By 2002, newspapers were reporting that the Maoists were launching attacks involving thousands of guerrillas, and the People's Army had announced that they had reached the level of fighting in units of brigade strength, of several hundred soldiers.
2 According to the official government structure: There are 4,000 VDCs (Village Development Committees) with ward chairperson/vice-chairperson who serve five-year terms; 28 municipalities; one metropolitan city (Kathmandu); four sub-metropolitan cities; 75 districts composed of VDCs and some municipalities; 14 zones; and five regions (Eastern, Middle, Western, Mid-West, Far West).
3 See photo 5.

5 REVOLUTIONARY WORK IN THE CITY

1 Many legal mass organizations that worked openly in support of the People's War have been forced to go underground. Especially after a State of Emergency was declared on November 23, 2001, it was impossible for such organizations to hold meetings, rallies, and other public events without being attacked by the police and having their members arrested.

6 GENERAL STRIKE IN KATHMANDU

1 *Kathmandu Post,* March 5, 1999.
2 This policy was written into the 'Common Minimum Policy and Programme of United Revolutionary People's Council, Nepal (URPC),' published in *The Worker #8,* January 2003.

CARRYING THE STORY FORWARD: THE PROBLEM OF DISINFORMATION

1 *Kathmandu Post,* August 10, 2002, '36 Journos Arrested at Different Locations.'
2 *Kathmandu Post,* September 13, 2002, 'Probe into Missing Journos' Case Assured.'
3 Amnesty International Report: 'Nepal: A Spiraling Human Rights Crisis,' April 4, 2002 at www.amnesty.org and the Committee to Protect Journalists, www.cpj.org
4 *Kathmandu Post,* July 24, 2002, 'Maoists Use Alcohol, Drugs Prior to Attack Army,' by Surya Chandra Basnet.

7 LAND IN THE MIDDLE

1 An article in *A World to Win* describes the class structure in the countryside: 'Nepalese rural society is divided essentially among landlords, rich peasants, middle and lower-middle peasants, poor peasants and landless peasants. Because of the semi-feudal relations of production, a lower-grade Nepali landlord may have barely as much wealth as a jobless European who simply enjoys reliable social security. A Nepali landlord generally is a person with more than enough land to survive, and his own house, cattle and fowl for fertilizer and food, and he does not himself work in his field or tend his cattle. A rich peasant is someone who has two yokes of oxen (4 oxen), cattle and fowl for milk and meat, as well as for fertiliser, and works in the field partly with hired labour and partly with shared labour (where people aid each other, particularly with seasonal work), and employs a servant in the house. Middle and lower-middle peasants are those who have one yoke of oxen, land enough to share labour, and production from the land is scarcely enough to eat for a year. The poor peasants are those who have a small plot of land with no cattle, and the production barely yields food for half the year or so. He does not share his labour but sells it. The landless peasants sell their labour all the time, and the life of the family amounts to a form of bonded slavery' ('Revolution in Nepal: A Better World's in Birth,' *A World to Win* #29, 2002).

2 An American program of DDT-spraying in the Terai greatly reduced the incidence of malaria and led to a large shift in population from the hills to the plains.

8 HOPE OF THE HOPELESS IN GORKHA

1 See photos 20, 25, 26, and 29.

CARRYING THE STORY FORWARD: REVOLUTIONARY POLICIES

1 Maoist Information Bulletin #6, document titled 'Let's Concentrate Total Force to Raise Preparations for the (Strategic) Offensive to a New Height Through Correct Handling of Contradictions.'

11 PEOPLE'S POWER IN ROLPA

1 See photo 13.

12 GUNS, DRUMS, AND KEYBOARDS

1 Mao Tsetung, 'Talks at the Yenan Forum on Literature and Art,' in *Selected Works of Mao Tsetung, Volume III* (Peking: Foreign Languages Press, 1967).

2 In 1967, a 'Spring Thunder' of revolutionary struggle broke out in Naxalbari, India. Led by Maoist revolutionaries, poor and landless

peasants, tea plantation workers, and tribal people in the northern part of West Bengal, near the border with Nepal, armed themselves with primitive weapons and guns and rose up against centuries of poverty and brutality.

CARRYING THE STORY FORWARD: CHILDREN IN THE WAR ZONE

1 *The Worker* #6, October 2000.
2 *State of the Rights of the Child in Nepal – 2003*, Child Workers in Nepal Concerned Center (CWIN) annual report, www.cwin-nepal.org/resources/reports.
3 *Kathmandu Post*, January 15, 2003, 'Int'l Community Has Paid Little Attention to Nepal: Human Right Watch,' by Akhilesh Upadhyay.
4 *Asia Times*, January 18, 2003, 'Nepal: Suffer the Little Children,' by Suman Pradhan.
5 *Kathmandu Post*, July 1, 2002, '30,000 Child Labourers in Nepal' and *Kathmandu Post*, July 6, 2002, '32,000 Children Working in 1,600 Stone Quarries: Report,' by Seema A. Adhikari.
6 *Kathmandu Post*, October 18, 2002, 'Fifty per cent Children Suffer from Malnutrition in Nepal.'

14 MARTYRS OF ROLPA

1 See photos 1 and 2.

15 FAMILIES OF MARTYRS: TURNING GRIEF INTO STRENGTH

1 See photo 6.
2 See photo 8.

CARRYING THE STORY FORWARD: THE RISING DEATH TOLL

1 *Kathmandu Post*, 20 December 2002, 'Unlawful Killings.'
2 *Kathmandu Post*, 21 December 2002, 'Unlawful Killings.'

16 WOMEN WARRIORS

1 See photo 31.
2 Rekha Sharma was later arrested along with her husband. She was subsequently released from jail and expelled from the Party and Women's Association for 'capitulating to the enemy.'
3 According to the *Kathmandu Post*. A woman dies every two hours in Nepal while giving birth due to the lack of safe maternity practices (*Kathmandu Post*, July 23, 2002).
4 On September 26, 2002, King Gyanendra approved The Country Code on the 11th Amendment Bill on property rights and abortion rights, which had been passed by the parliament. This meant that for the first time in

Nepal's history, unmarried women have rights to their parental property and married women have rights to their husband's property. Married women are now supposed to have a share of the husband's property; the same pertains to divorced women; widows have full ownership of their husband's property. Before the passage of this bill, property rights were only granted to women above 35 years of age who remain unmarried. The bill also granted women the right to an abortion: up to 12 weeks of pregnancy, up to 18 weeks in cases of rape and incest, and at any time if there is risk to the woman's health or if the fetus is impaired. Women's rights groups in Nepal are critical of the amendment, saying all it does is take away the 35-year age restriction for unmarried women to claim their share of parental property – and a woman who gets married still has to return the property to her brothers. Women's rights groups have also pointed out that the law does not address the rights of the hundreds of women who are currently in prison on charges of abortion and infanticide. It has been reported that women have continued to be convicted and jailed for having an abortion even after the passage of the bill (*Kathmandu Post*, June 5, 2003, 'If Abortion is Legal, Why are These Women in Prison?' by Sudha Shrestha). It is also the case that many women in Nepal won't have access to abortions because of the Bush administration's global gag rule, which prohibits non-governmental organizations receiving US assistance from performing abortions or making necessary referrals.

17 NEW WOMEN, NEW PEOPLE'S POWER

1 See photos 14, 27, and 28.
2 See photo 24.
3 See chapter 16, note 4 on the passage of the 11th Amendment Bill giving women new rights to land and abortion.
4 The Common Minimum Policy and Programme of the United Revolutionary People's Council, Nepal, adopted by the First National Convention of the Revolutionary United Front under the Leadership of CPN (Maoist) in September 2001 states, under SECTION-IX, Women and Family: 'If both husband and wife desire and request for, divorce may be granted. If only one party demands for it, the People's Government shall reserve the right whether to or not to allow for the divorce, after necessary investigation. Taking into account the traditional oppression of women, a more sympathetic consideration should be given to women in such disputes. After the divorce, common property shall be divided on equal basis, but as far as the taking care of children is concerned, the father shall bear 2/3 and the mother 1/3 responsibility (*The Worker* #8, January 2003).

18 MAGAR LIBERATION

1 The policy of the CPN (Maoist) is to fight for a countrywide united government of anti-feudal and anti-imperialist patriotic forces, within

which the oppressed nationalities will have the right of self determination. *A World to Win News Service* reported that, on January 9, 2004, a Magar Autonomous Region was announced at a gathering of 75,000 people of that region and a convention of 130 representatives of the Magar Autonomous Region held on January 7–8 elected a 27-member Working Committee. The CPN (Maoist)'s Krishna Sen News Service reported at this time that other oppressed nationality regions would also be declaring regional and national autonomy in different parts of the country.

21 CAMPING WITH THE PEOPLE'S ARMY

1 See photo 9.
2 Platoon Commander Sundar was later promoted to Regional Bureau of the Party, but was killed in 2001 in what the Party reported as a 'fake encounter with the enemy army.'
3 See photo 17.

22 RED SALUTE IN THE WEST

1 In November 2001, for the first time, King Gyanendra fully unleashed the Royal Nepal Army against the Maoist insurgency.

AFTERWORD

1 *South Asia Intelligence Review*, Volume 2, No. 21, December 8, 2003 points out: 'The withdrawal of the agencies of the state is complemented by the widening sphere of Maoist presence and activity. Towards the latter half of 1999, 20 of Nepal's districts were considered "seriously affected" by Maoist violence. By 2001, their number was up to 68. Now, all 75 districts in the country, including the capital, Kathmandu, have witnessed significant Maoist violence.'
2 See Mao, *Selected Military Writings*. In actuality, at least by 2003, the numbers were not literally '10 to 1' against the Maoists, but this basic point of military theory still applies, especially in terms of things like more advanced weaponry, foreign military assistance, etc.
3 *South Asia Intelligence Review*, Volume 2, No. 21, December 8, 2003 points out that in Rolpa, 'in 1996, there were 33 police stations, with the largest manned by 75 men, but most of the others with a strength less than 20. When the post at Ghartigaun in western Rolpa was attacked in 1999, for example, it had a complement of 19. Fifteen were killed, the others wounded; the station was totally destroyed and was never re-garrisoned. In 1998, two such stations were abandoned; in 1999, a further 16 (including Ghartigaun); in 2000, six more; in 2001, another four; and in 2002, three – leaving a total of just two police stations for the entire population of nearly 211,000.'
4 Mao Tsetung's military writings talk about the continued role of guerrilla warfare and smaller military units, even at later stages of protracted people's war (strategic equilibrium and strategic offensive). Since the

resumption of fighting between the PLA and government forces at the end of 2003, the CPN (Maoist) has discussed the need, for various reasons, to give some emphasis to 'less concentrated' actions, as part of their overall strategy.

5 *Kathmandu Post*, January 17, 2003, 'Int'l Community Has Paid Little Attention to Nepal: Human Right Watch,' by Akhilesh Upadhyay.
6 *Kathmandu Post*, March 5, 2002.
7 Arjun Karki and David Seddon, 'The People's War in Historical Context,' in Arjun Karki and David Seddon, eds., *The People's War in Nepal, Left Perspectives*, (New Delhi: Adroit Publishers, 2003), p. 42.
8 Statement from the British Department for International Development, 'UK Hosts an International Conference on Nepal,' June 20, 2002.
9 Karki and Seddon, 'The People's War in Historical Context,' p. 42.
10 *Kathmandu Post*, 'India Extradites Six Hospitalised "Maoist Rebels",' December 20, 2002.
11 *Kathmandu Post*, 'Govt Urges Rebels to Sit for Peace Talks,' December 1, 2002.
12 Bhagirath Yogi, 'Analysis: What Next for Nepal?,' BBC Nepali Service, September 4, 2003.
13 *A World to Win News Service*, January 19, 2004.
14 Amnesty International report at web.amnesty.org/web/wire.nsf/November2003/Nepal
15 *Kathmandu Post*, 'US and British Envoys Suggest Reconciliation,' September 1, 2003.
16 Worldview, Chicago Public Radio, WBEZ, Jerome McDonnell interview with Michael Malinowski, November 28, 2003, Chicago.
17 See, for example, 'Prairie Fire from Nepal' by B. Raman, former Additional Secretary, Cabinet Secretariat, Government of India, Financial Daily from THE HINDU group of publications, 23 July 2001 and 'Maoist Incursions across Open Borders' by P.G. Rajamohan in *South Asia Intelligence Review*, Volume 2, No. 22, December 15, 2003, and 'MCC and Maoists: Expanding Naxal Violence in Bihar' by Sanjay K. Jha and other articles on the Institute for Conflict Management website.
18 *Kathmandu Post*, June 12, 2002.
19 Rita Manchanda, '"War for Peace"Approach Promises More Bloodshed' (InterPress Service, December 26, 2002).
20 Bob Avakian, 'Two Humps in the World Revolution: Putting the Enemy on the Run,' *Revolutionary Worker*, January 18, 1998.
21 See 'International Communist Movement and Its Historical Lessons,' Document of the Second National Conference of the Communist Party of Nepal (Maoist), reprinted in *A World to Win* #27, 2001 and *Himalayan Thunder*, the quarterly bulletin of the CPN (Maoist), May 2001. This document from the CPN (Maoist) states: 'Due to the uniqueness of the economic, political, cultural and geographical conditions and the unchallenged hold of Indian monopoly capitalism, it will be very difficult for any single country of this region to successfully complete the new national-democratic revolution and; even if it succeeds following the distinct contradictions, it will be almost impossible for it to survive. The revolutionaries need to seriously concentrate on the fact that a particular

country, or a particular territory of a country, shall be liberated through the force of the common and joint struggle of the people of this region following the unequal stage of development, and that it can play only a particular role of base area for the revolution in the whole region.'

22 Onesto, 'Red Flag Flying on the Roof of the World.'

23 Press statement of the Coordination Committee of Maoist Parties and Organizations of South Asia (CCOMPOSA), 1 July 2001, published in *People's March* (Vol. 2, No. 9, September 2001). Available on the web at www.peoplesmarch.com

References

A World to Win, 'Revolution in Nepal: A Better World's in Birth' (#29, 2002).

Avakian, Bob. *Conquer the World? The International Proletariat Must and Will*, in *Revolution* magazine (Special Issue, #50, 1981).

—— 'Democracy: More Than Ever We Can and Must Do Better Than That,' in *A World To Win* (#17, 1992).

—— 'Getting over the Hump,' a series of articles published in the *Revolutionary Worker* and available online at rwor.org/chair_e.htm

—— *Mao Tsetung's Immortal Contributions* (Chicago: RCP Publications, 1979).

Aziz, Barbara Nimri. *Heir to a Silent Song, Two Rebel Women of Nepal* (Kathmandu: Centre for Nepal and Asian Studies, Tribuvan University, 2001).

Banerjee, Sumanta. *India's Simmering Revolution, The Naxalite Uprising* (London: Zed Books, Ltd, 1980).

Bhattarai, Baburam. *The Nature of Underdevelopment and Regional Structure of Nepal, a Marxist Analysis* (Delhi: Adroit Publishers, 2003).

Dixit, Kanak Mani and Shastri Ramachandaran, eds. *State of Nepal* (Nepal: Himal Books, 2002).

Hoftun, Martin, William Raeper, and John Whelpton. *People, Politics & Ideology, Democracy and Social Change in Nepal* (Kathmandu: Mandala Book Point, 1999).

Karki, Arjun and David Seddon, eds. *The People's War in Nepal, Left Perspectives* (Delhi: Adroit Publishers, 2003).

Mao Tsetung. *Selected Military Writings* (Peking: Foreign Languages Press, 1968).

—— *Talks at the Yenan Forum on Literature and Art*, in *Selected Works of Mao Tsetung*, Volume III (Peking: Foreign Languages Press, 1967).

Mikesell, Stephen Lawrence. *Class, State, and Struggle in Nepal* (New Delhi: Manohar, 1999).

Ogura, Kiyoko. *Kathmandu Spring, The People's Movement of 1990* (Nepal: Himal Books, 2001).

Paskal, Anna. *The Water Gods: The Inside Story of a World Bank Project in Nepal* (Montreal: Véhicule, 2000).

Shrestha, Aditya M. *Bleeding Mountains of Nepal, A Story of Corruption, Greed, Misuse of Power and Resources* (Kathmandu: Ekta Books, 1999).

Verma, Anand Swaroop. *Maoist Movement in Nepal* (New Delhi: Samkalin Teesari Duniya, 2001).

WEBSITES FOR INFORMATION ABOUT THE MAOIST PEOPLE'S WAR IN NEPAL

A World To Win – awtw.org
Communist Party of Nepal (Maoist) – cpnm.org
International Nepal Solidarity Forum – insof.org
Revolutionary Worker/Obrero Revolucionario – rwor.org
Li Onesto website – lionesto.net

Index